AF878691

PHENOMENOLOGY AND DIALECTICAL MATERIALISM

BOSTON STUDIES IN THE PHILOSOPHY OF SCIENCE

EDITED BY ROBERT S. COHEN AND MARX W. WARTOFSKY

VOLUME 49

TRÂN DUC THAO

PHENOMENOLOGY AND DIALECTICAL MATERIALISM

Translated by

Daniel J. Herman *and* Donald V. Morano

Edited by

Robert S. Cohen

D. REIDEL PUBLISHING COMPANY

A MEMBER OF THE KLUWER ACADEMIC PUBLISHERS GROUP

DORDRECHT / BOSTON / LANCASTER / TOKYO

Library of Congress Cataloging in Publication Data

Trân, Duc Thao.
Phenomenology and dialectical materialism.

(Boston studies in the philosophy of science ; v. 49)
Translation of: Phénoménologie et matérialisme dialectique.
Bibliography: p.
Includes index.
1. Phenomenology. 2. Dialectical materialsm. 3. Husserl, Edmund, 1859–1938. I. Cohen, Robert Sonné. II. Title. III. Series.
Q174.B67 vol. 49 [B829.5] 001'.01 s [146'.32] 85–35
ISBN 90-277-0737-5

Published by D. Reidel Publishing Company,
P.O. Box 17, 3300 AA Dordrecht, Holland.

Sold and distributed in the U.S.A. and Canada
by Kluwer Academic Publishers,
160 Old Derby Street, Hingham, MA 02043, U.S.A.

In all other countries, sold and distributed
by Kluwer Academic Publishers Group,
P.O. Box 322, 3300 AH Dordrecht, Holland.

Translated from Trân Duc Thao's *Phénoménologie et matérialisme dialectique* (Paris: Minh Tan, 1951; re-issued New York: Gordon & Breach Science Pubs., Inc., 1971).

All Rights Reserved.
© 1986 by D. Reidel Publishing Company.
No part of the material protected by this copyright notice may be reproduced or utilized in any form or by any means, electronic or mechanical, including photocopying, recording or by any information storage and retrieval system, without written permission from the copyright owner.

Printed in The Netherlands.

TABLE OF CONTENTS

EDITORIAL PREFACE

Trân Duc Thao, a brilliant student of philosophy at the École Normale Supérieure within the post-1935 decade of political disaster, born in Vietnam shortly after the First World War, recipient of a scholarship in Paris in 1935–37, was early noted for his independent and original mind. While the 1930s twisted down to the defeat of the Spanish Republic, the compromise with German Fascism at Munich, and the start of the Second World War, and while the 1940s began with hypocritical stability at the Western Front followed by the defeat of France, and the occupation of Paris by the German power together with French collaborators, and then ended with liberation and a search for a new understanding of human situations, the young Thao was deeply immersed in the classical works of European philosophy. He was also the attentive but critical student of a quite special generation of French metaphysicians and social philosophers: Gaston Berger, Maurice Merleau-Ponty, Emile Bréhier, Henri Lefebvre, René le Senne, Jean-Paul Sartre, perhaps the young Louis Althusser. They, in their several modes of response, had been meditating for more than a decade on the philosophy of Edmund Husserl, which came to France in the thirties as a new metaphysical enlightenment – phenomenology. With Husserl's phenomenology, there also came the powerful influences of a revived Hegel (of the *Phenomenology*) and of Martin Heidegger's existentialism, and, in a tangle of variants, there came a startling renewed investigation of Marx. The young Trân Duc Thao joined the search for objective truth, worked to overcome both psychologism and every weakening of knowledge by subjectivist limitation, investigated Husserl's writings in print and in the fine archives at Louvain (with the kindly help of H. L. van Breda). His progress was dialectical, Socratic and Hegelian, but also it was a material dialectic due both to his Marxist studies and to the grim tasks of the greater liberation in his social life-world – the liberation of Vietnam.

Thao's themes drove him to the border of Husserl's thought, just as Thao saw Husserl himself driven toward the apparent relativism of the final *Krisis* manuscripts. The privileged, indeed most precious, phenomenological activity is that of 'constitution', for which there is the endless work of passing from naive certainty to the developed no-longer-naive certainties of intentional praxis. Such praxis has to be understood in so many ways of becoming, and especially of becoming aware; the awareness is of what is mine and what is the historical world; the praxes are social, political, symbolic, communicative, even philosophical. If to constitute is to act, the phenomenological re-establishment of meaning must investigate the temporal within action, both as the particular quality of a time-consciousness and as the dialectic of past, present and future: the past made present cannot be mine alone; the future yet to be chosen, 'constituted', must be inter-subjective and historical, so it is Husserl's future as 'horizon'. For Trân Duc Thao, the human activity of 'constitution' in the present demands understanding, which retrospectively deals with genesis and prospectively with goals. Genetic understanding entails materialist science, the investigation of the evolutionary biological foundation and historical development of consciousness, and especially of that practical consciousness which is language. Understanding of goal-directed present 'constitutive' action suggests a materialist science of intentional possibilities, a political economy of historically determined needs, desires, ideologies, resources and alternatives.

Whether the phenomenological method is capable of leading to such understanding, whether that method can situate the social sciences so that they may rightly formulate their investigations, is shown to be doubtful in Part One of this book. Thao here, and in his articles of those same early years, sets aside the existentialist turn (effectively close to Lukacs's 1947 critique of the existentialist pessimism of 'dreadful freedom'); for Thao, the way forward must lead through the theory and practice of Marxism. What Marxism may say to the philosopher at this point was roundly debated in France in the two decades after Liberation. For Trân Duc Thao, the Marxist heuristic led to the question of clarifying the nature of production. Thus production is the moment of constitution (and in Thao's later work we see how re-presentation, gesture, and the embryology of language elaborates production at the origin of human mentality). But Husserl, as Thao tells us in his careful exposition, provided only a program: to go beyond the undoubted virtues of anti-formalism and anti-dualism, phenomenology would have to transcend its own analytic restrictions and enter upon analysis of the historical materialism of the subject, a critique of the forces pressing upon the subject as that subject's very life-world. Thao writes of 'the historical movement of reality' and at the conclusion of his Husserlian study he summarizes: ". . . there is no longer any valid reason to refuse to (the) constituting subjectivity its predicates regarding reality". And why is this possible, why can we transcend the transcendental idealism? Because ". . . it is *nature* itself *becoming-subject*" (129).

In Part Two, Thao sketched his alternative to the 'radical contingency' of Husserl's *Weltkonstitution* (130). He wrestled then, and since (and how many others have done likewise), to argue, indeed to *show*, the plausibility of a non-reductive materialism. His epigraph from Engels set the problem: "Matter is not a product of mind, but mind itself is merely the highest product of matter". The sketch was supported by Thao's persuasive interpretation (1948) of Hegel's 'Phenomenology' as naturalistic, anti-reductionist, indeed nature becoming mental and substance becoming subject. The material content of Hegel's logic and phenomenology receives much deeper investigation in Thao's major essay on the Hegelian dialectic (first published in the Vietnamese language in Hanoi (1956), then in French (1965)). Here Thao elaborates the forms of work (social praxis) and of consciousness within historical stages, typically Hegel on desire, master and slave, death and the struggle for life, private truth and public intersubjectivity. But beyond his materialist reading of Hegel was the need for scientific investigations. Thao's essays of the 60s and early 70s, and his monograph of *Investigations into the Origins of Language and Consciousness,* are scientific works, however much they are also philosophical, i.e. explorations of a historical materialist dialectic. If Thao's Marxist exploration of the limiting crisis within phenomenology is compared with his phenomenological reading of Marxist philosophy (as in 'The Dialectic of Human Societies as the Becoming of Reason', Part Two, Chapter 2, below, but also in his early interpretation of Vietnamese practical affairs in *Les Temps modernes,* 1947), we see that his turn to scientific inquiries was the inevitable and natural development. His French critics, with respect for his sensitive and acute qualities, pointed to the *philosophical* (they often wrote 'theoretical') limitation of Thao's Marxist alternative; they were less ready to accept that Trân Duc Thao and a sober Marxism could overcome this limitation through the sciences, through Soviet as well as Western anthropology, archeology, paleontology, linguistics, cognitive and developmental psychology, historical sociology. We see the young Trân Duc Thao's courageous and complex argument in this book – concerning Husserl, Hegel and Marx – as a classical text of 20th-century philosophy, and his subsequent synthesis of work in the human sciences as his mature and profoundly stimulating accomplishment in the philosophy of science.

* * *

Works of Trân Duc Thao in Western languages

1943: 'The phenomenological method in Husserl' (Thesis, in French, unpublished).

1946: 'Marxisme et phénoménologie', *La Revue Internationale* 2, 168–174.

1947: 'Les relations Franco-Vietnamiennes', *Les Temps modernes* 2, 1053–1067.

'Sur l'interprétation trotzkyiste des événements d'Indochine', *Les Temps modernes* 2, 1697–1705.

1948: 'La "Phénoménologie de l'esprit" et son contenu réel', *Les Temps modernes* **3**, 492–519.

1949: 'Existentialisme et matérialisme dialectique', *Revue de Métaphysique et de Morale* **54**, 317–329.

1950: 'Les origines de la réduction phénoménologique chez Husserl', *Deucalion* **3**, 128–142.

1951: *Phénoménologie et matérialisme dialectique*, Éditions Minh-Tân, Paris. 368 pp.

1965: 'Le "Noyau rationnel" dans la dialectique hegelienne', *La Pensée* No. 19, 3–23. (First published in 1956 in the Vietnamese journal *Tâp san Dai hoc* at the University of Hanoi.)

1966: 'Le mouvement de l'indication comme forme originaire de la conscience', *La Pensée* No. 128, 3–24.

1969: 'Du geste de l'index a l'image typique', Part I, *La Pensée* No. 147, 3–46.

'Du geste de l'index a l'image typique', Part II, *La Pensée* No. 148, 71–111.

1970: 'Du geste de l'index a l'image typique', Part III, *La Pensée* No. 149, 93–106.

Fenomenologia e materialismo dialettico, Italian translation by Roberta Tomassini, Lampugnani Nigri Editore, Milan, xix and 282 pp. With a bio-bibliographical note by the translator and a critical introduction by Pier Aldo Rovatti (to which we are indebted).

1973: *(Recherches sur l'origine du langage et de la conscience.* Éditions Sociales, Paris. 343 pp.

1984: *Investigations into the Origin of Language and Consciousness* (Boston Studies in the Philosophy of Science 44). English translation by Daniel J. Herman and Robert L. Armstrong; edited by Carolyn Fawcett and Robert S. Cohen, D. Reidel, Dordrecht and Boston. xi and 214 pp.

1985: *Phenomenology and Dialectical Materialism* (Boston Studies in the Philosophy of Science 49). English translation by Daniel J. Herman and Donald V. Morano; edited by Robert S. Cohen, D. Reidel, Dordrecht, and Boston. xxviii and 242 pp.

Note: A sampling of critical appreciations of this book and the essays of Trán Duc Thao may be found in Paul Ricoeur 'Sur la phenomenologie', *Esprit* **21**, 827–836 (1953); the A. De Waelhens review in *Critique* **58**, 85–88 (1953) and in *Rev. Metaph. et morale* **58**, 310–312 (1953); G. Neri in his *Prassi e conoscenza*, 149–163 (Feltrinelli, Milan, 1966); Roger Garaudy in his *Perspectives de l'homme*, 304–313 (PUF, Paris, 1969); P. Lyotard in his *La Phénoménologie*, 110–126 (PUF, Paris, 1967). See also Michael Kelly's perceptive exposition of Thao's 'materialist inversion' in the 'rational kernel' paper of 1965 in Kelly, *Modern French Marxism*, 160–164 (Johns Hopkins University Press, Baltimore, and Blackwell, Oxford, 1982); and the remark by J. Derrida in *Philosophy in France Today*, edited by Alan Montefiore p. 38 (Cambridge University Press, 1983).

On the translation: Where Husserl uses 'Abschattungen', Thao uses 'silhouettes' as do Thao's translators, while Husserl translators generally use 'perspective variation'. For Husserl's 'Erlebnis', Thao uses 'vecu' and Thao's translators use 'lived experience' or 'lived' depending upon syntactical need, while Husserl translators use 'experience'. For Husserl's use of 'ursprünglich' and similar 'ur-' prefixed terms, Thao uses 'originaire', his translators use 'primordial' and the Husserl translators use 'original' or 'primary' or 'primal'.

Footnote numbers followed by asterisks indicate notes introduced by the translators as required by the French text.

* * *

We are grateful to Daniel J. Herman and Donald V. Morano for the linguistic and philosophical skills which they gave so generously to their craftsmanlike translation.

November 1985

ROBERT S. COHEN
Boston University

MARX W. WARTOFSKY
Baruch College, City University of New York

TRANSLATORS' FOREWORD

In the second half of the twentieth century, when communication and dialogue between all peoples of the world is so urgently needed, this philosophical work of 1951 by the Vietnamese Trân Duc Thao merits our attention.

Professor Thao received much of his education in the West. In his twenties he was a scholarship student at the prestigious École Normale Supérieure of Paris, at which, from his first days, he was regarded by his instructors as 'prodigiously talented'. His studies there spanned the years of the German occupation. Though never studying under or even meeting Edmund Husserl, he made several visits to the Husserlian archives at Louvain and was provided by Fr. Van Breda with unpublished transcriptions of Group C of Husserl on temporality. He alludes to these rather extensively in *Phenomenology and Dialectical Materialism*. He has remained in Vietnam since 1950, serving between 1958 and 1964 as an adviser and agricultural minister and later as the director of a school. Therefore, his extensive and varied background, in addition to the timeliness of the topic, *Phenomenology and Dialectical Materialism*, leads us to pay serious attention to this 1951 study. For phenomenology and Marxism are certainly two of the major philosophical movements – if not the most significant – of this century.

The Preface to this book was written last and should be read with great care since it vividly, if not sometimes rhetorically and even tendentiously, articulates the thesis of this study. Thao writes: "It is not a question in any sense of a mere juxtaposition of two contradictory points of view: Marxism appears to us as the only conceivable solution to problems raised by phenomenology itself."

Thao's rejection of phenomenology is based on an internal contradiction found in Husserl's own work – particularly the later work – since the practical results of phenomenology are incompatible with the theoretical principles from which these concrete analyses were elaborated. Husserl's undertaking from the first philosophy of essences to 'The Origins of Geometry' evolves within the horizon of idealism which is dominated with the 'constitution of the thing'. This constitution, as well as the constitution of all meaning and truth based on the phenomenological reduction which introduces this constitution reduces the being of the thing to a constituted being. But, as

Thao rightly points out, according to Husserl himself, all constitution, whether of the thing, being, values or persons, is always founded on an antepredicative perception; the categorial intuition, or the predicates of a proposition, are always formed on the background of the world, with the consequence that the constitution of the world is no longer the product of the transcendental Ego, but the product of each man's consciousness, of each man's individual experiences in his own particular environment. According to Thao, not only does phenomenology end up in a total skeptical relativism: "the merchant at the market has his own market-truth," it also uselessly duplicates empiricism for "the concept 'transcendental' was superfluous from the outset, since it maintains a strict identity of content between 'pure consciousness' and natural consciousness." Thao praises Husserl for appreciating the fact that "the universal is constituted in the real movement of time" for "the principal merit of phenomenology was its definitive destruction of formalism in the very horizon of idealism and its placing of all problems of value on the level of the concrete." Nevertheless, he concludes that "the great problem of our time . . . has found its solution for a long time in the Marxist dialectic, which defines the only valid process for a constitution of lived significations on the foundation of material reality."

In an elegant, laconic, and remarkably lucid prose, Thao elaborates the rationale that motivated Husserl's philosophizing: Husserl wished to go beyond psychologism in order to justify and delineate the realm of objective truth. But because of his idealistic scaffolding – the transcendental Ego, the phenomenological reduction, the constitution, etc., and being a bourgeois philosopher – Husserl was unable to achieve his original project and to realize fully that man's labor is the only source of truth.

Thao's knowledge of Husserl, as well as the entire history of philosophy, is most impressive, and he has the ability to elucidate and bring to life some of the most abstruse epistemological writings of Kant, Hegel, Descartes, etc. It would be foolhardy and redundant for us to attempt to summarize this work: we shall content ourselves with highlighting examples.

Both the beginner and the advanced student of phenomenology can delight in the lucidity of Thao's explanation of the eidetic reduction, the necessity of taking into account the subjective pole in every act of knowledge, and the difference between a causal and a phenomenological explanation. Thao provides for an especially illuminating discussion of the problem of error. He makes the paradoxical assertion: "The apodicticity of self-evidence does not militate against the correction of error." But, *prima facie*, it would seem as if apodicticity would preclude revision. Yet, Thao argues rather convincingly:

> Truth, being defined in terms of self-evidence, cannot by the very fact be posited in the absoluteness of an in-itself, for it belongs to the precise meaning of all self-evidence to be fallible. Self-evidence exists as such only in its actual lived experience, which is renewed at each moment. Such a movement implies an absolute privilege of present actuality over the results previously acquired and the right of constantly rectifying them. Thus, truth is defined only in terms of its becoming, which must not be understood as an intelligible movement of ideas, but as an actually lived temporality.

However, lest these remarks be construed as maintaining epistemological relativism in the long haul, Thao insists upon an irrefragable grounding for objective knowledge: "Evidence can be corrected or annulled by other evidence. But such a movement precisely presupposes the absolute right of evidence in general."

The first part, that is, the phenomenological part of this work, is a treasure house of such insights. Then in Part Two, Thao attempts to disclose dialectical materialism as the logical issue of a relentlessly pursued phenomenology. He traces the deplorable mathematizing of nature that resulted because of Descartes' pernicious dichotomy between a thinking mind and an extended object. "Scientific objectivity was established precisely through the radical elimination of all subjective signification. After Descartes, metaphysical rationalism gave itself the task of radicalizing the dualism and suppressing the mystery of the union in man by referring him to God, conceived as absolute substance or as the source of harmony." Furthermore, this abstract view of matter, in reducing the principle of identity or non-contradiction to an abstract mathematical point, failed to take into account the fact that "concrete identity includes difference and change." But, Thao argues, the dialectical aspect of nature that had been noted by Engels "has recently been shown in detail by natural science."

Having vitalized matter, Thao proceeds to explain the emergence of human consciousness and language by means of a movement within nature itself. He criticizes the Neoplatonism of phenomenology in its characterization of mind as ontologically prior to, and independent of, matter. Rather, Thao accounts for the emergence of consciousness as "the movement of deferred behavior, which sends us back to deferred circuits in the course of the nervous influx." Consequently, "subjectivity is but the formal aspect of the real dialectical process in which each new structure represses the one that precedes it and absorbs it in a lived interiority." "The sensorial impression, as the primordial form of lived experience . . . is only the irritability of the cellular element absorbed and repressed in the reaction of the whole organism." "The development of the nervous system is presented as a dialectic in which the successive levels of behavior are superimposed on each other – each formation inhibiting the preceding one and absorbing its influxes in a broader regulation in which they are maintained as suppressed, preserved, transcended."

Critics of Thao's work, such as Paul Ricoeur[1] and Jean-Paul Sartre,[2] though favorably disposed toward Thao's grounding of philosophical investigation in concrete experience, ultimately reject his account of the origins of consciousness and, especially, self-consciousness. For they argue: how, possibly, could matter give rise to the *idea* of matter?

Thao himself, dissatisfied with the second part of *Phenomenology and Dialectical Materialism* wrote essays in the 1960s and early 70s which develop and modify his views regarding the origins and nature of language and consciousness. These essays which form his second major work, *Recherches sur l'origine du langage et de la conscience,*[3] are based on Marx's classical theory which considers that consciousness is from the outset a social product, elaborated in the material activity and material relations of men, in the language of real life. Such a conception, which is strictly materialistic, allows for the description of lived experience on an entirely objective and scientific plane, which definitively eliminates, by making superfluous, the phenomenological method from which the author could not entirely free himself in *Phenomenology and Dialectical Materialism*.

However, though Thao may have modified and even rejected some of his views regarding phenomenology and dialectical materialism in his most recent work, it is still useful and most instructive to study *Phenomenology and Dialectical Materialism*, which has had a wide influence on the European continent these past three decades, since we find in this work a serious philosopher grappling with the relationship between phenomenology and dialectical materialism: for whoever would come to serious grips with contemporary philosophy must take into account these two movements.

DANIEL J. HERMAN
The University of West Florida

DONALD V. MORANO
Los Angeles, California

NOTES

[1] His review-essay, 'Phenomenology' first appeared in *Esprit* (1953) **21**, pp. 821–838. Daniel J. Herman and Donald V. Morano translated it into English for the Fall 1974 Husserl issue of the *Southwestern Journal of Philosophy*.

[2] 'Materialism and Revolution' (1949), in *Philosophy in the Twentieth Century*, Vol. 3, edited by William Barrett and Henry D. Aiken, New York (1962), p. 389.

[3] Trân Duc Thao: 1973, *Recherches sur l'origine du langage et de la conscience*, Paris: Editions Sociales. The English translation of this work is published as *Boston Studies in the Philosophy of Science*, Vol. 44 (1984).

ACKNOWLEDGMENTS

The translators wish to thank Professors Herbert Spiegelberg and Paul Ricoeur for preliminary assistance in suggesting knowledgeable individuals and sources in regard to Trân Duc Thao. Ms Roberta Tomassini, the translator of this work into Italian as *Fenomenologia e materialismo dialettico* Milano, Lampugnani Nigri, 1970, deserves warm thanks for her many acts of kindness to us, providing us with a copy of her translation and her dissertation on Thao and suggesting other individuals who might be able to further assist us in our project. We thank Fr. Herman Van Breda, the curator of the Husserl Archives of Louvain for providing us with a vivid account of biographical details regarding Thao and for his suggestions and helpfulness to us in our project.

Michael Smith, Morano's teaching assistant at Loyola University during the academic year of 1971–72, deserves thanks for his careful and constructively critical review of the first third of this translation. Surely, the translation is more accurate and fluent because of his suggestions.

Mrs Marianne McDaniels of Pensacola, the secretary-typist of the philosophy department of the University of West Florida, has kept this project a viable one, conscientiously typing and retyping the many versions and revisions that the finicky collaborators, some 1000 miles apart, submitted to her. Surely Daniel Herman on the scene, more than Donald Morano, has appreciated the patience and good cheer and unflagging devotion of Marianne. But Morano himself can well realize that without her efforts this project might never have been completed.

Donald Morano, wishes to extend special appreciation to his first French teacher, Miss Hazel Writer. She bore with his atrocious pronunciation for three years at Grafton Street Junior High School, Worcester, Massachusetts. Therefore, if it were appropriate for a translator to dedicate his translation to someone, he would certainly dedicate his share of this translation to his devoted French Teacher, Hazel Writer.

Professor Morano acknowledges with gratitude a grant from Loyola University to defray some of the clerical and travel expenses resulting from this project. We both wish to thank the Matchette Foundation for the grant it has contributed to us for translating this work.

Lastly, Daniel Herman wishes to heartily thank Anne who in her own way made this work possible by keeping little Nicole busy with her toys rather than with her daddy's translation.

AUTHOR'S PREFACE

The work that we present to the public consists of research belonging to different times and inspirations. In the first part, written between 1942 and 1950, we set forth the essential features of phenomenology from a purely historical point of view and in the perspective of Husserl's own thought. Our critical objections serve only to make evident internal contradictions found within the Husserlian corpus itself. In contrast, the second part, completed in 1951, is situated entirely within the position of dialectical materialism. It is true that there we take up again certain technical results of lived analyses, but only in terms of pure positive data, completely freed from the philosophical horizon that dominated Husserl's descriptive method. However, it is not a question in any sense of a mere juxtaposition of two contradictory points of view: Marxism appears to us as the only conceivable solution to problems raised by phenomenology itself.

Our task in setting forth Husserl's thought was a relatively easy one, since it was concerned only with the *theory* of phenomenological analysis under the three aspects that appeared successively in its evolution: the description of essences, the static explication of lived experience [*vécu*], and finally a genetic explication. Its concepts were simple enough, and, in addition, amply developed in the published works. But, obviously, theory is worthless without practice, and for a long time we believed that within the very presentation of the method should be included the achieved results of the method; however, the most important part of this work has remained unpublished.[1] It is here that we have encountered extraordinary difficulties, which are responsible for the long delay in the completion of this work and have radically reversed its orientation.

The examination of unpublished manuscripts demonstrated, in fact, that the concrete analyses took a direction that was incompatible with the theoretical principles from which these concrete analyses were elaborated. From the beginning of our study of Husserl (in a work written in 1942 of which we present here only the first chapter), we had surmised the contradiction because of certain enigmatic developments within the published works. However, we thought that we would be able to resolve this contradiction by a simple broadening of our perspective, which would remain faithful to the essential phenomenological inspiration. But, after long hesitation we found that, on account of the actual descriptions that abound in the manuscripts, we had to renounce once and for all any hope of reconciling the *concept* of phenomenology with its actual *achievement*. Since then, there has no longer been any question of remaining within the limits of a historical study, even with all the liberties which are permitted to the interpreter in the field of the history of philosophy. The enumeration of concrete analyses (which we give in the last chapter of Part One) involves a total and definitive break with the fundamental principles of the doctrine.

Since the *Logical Investigations* (1900–1901), every predication presupposes an antepredicative perception, not simply as an antecedent of the fact, but as a condition of truth. The *Ideas* (1913) adds that the sensible reality, so perceived, is the original resting place on which are constituted values and ends and, by the same token, all the cultural formations of 'objective spirit'. The last works systematize this point of view in presenting the life-world (*Lebenswelt*) as the origin and foundation for every intelligible signification – the world to which it is necessary to return constantly as the living source of all truth. The life-world is revealed as the domain of human history. The *genetic* method indicated in *Formal and Transcendental Logic* (1929) and *Cartesian Meditations* (brought out the same year) is developed in *Crisis of European Sciences* (1936) and the work on the *Origin of Geometry* (1939) under the form of a *historico-intentional* analysis, where the universal is constituted in the real movement of time. Since then, phenomenological idealism found itself superseded by the method of lived analysis: The genesis of the world in absolute consciousness has become confused with the actual becoming of real history, and the doctirne of the transcendental Ego has appeared only as a stylistic expression that conceals in philosophical terminology the creative value of human labor.

As evident as these consequences had appeared to us since our study of 1942, we still could not forget the fundamental phenomenological theme of the critique of psychologism. Unless one would be willing to fall back into the contradictions of sceptical relativism, the historicity of the Ego would have to be interpreted as the actuality of the eternal. But it does not seem that the notion of a constituent genesis of temporality can be accepted except under the aegis of an essence of temporality that is itself atemporal. The study of the unpublished works permits us to do away with every illusion in this regard. The *Weltkonstitution* is revealed here as resting totally on the sensible data (i.e., the kinesthetic and sensible configuations) just as they are constituted on a level proper to animals. There is such a visible acceptance of the primordial world with its elementary sensory-motor coordinates that there can be no doubt that the transcendental subject thematized by phenomenology must not be identified, strictly speaking, with the man of flesh and blood who is evolving in the real world.

Since that time one should take seriously the exceptional importance that Husserl has consistently granted in phenomenology to the 'thing' (*Ding*). The intersubjective communities and the spiritual entities that are constituted therein are firmly grounded in natural psychic realities that, in their turn, are grounded in physical realities. "Finally, at the base of all other realities one finds the natural reality, and so the *phenomenology of material nature*, undoubtedly, occupies a *privileged position*." [2] If we remind ourselves that psychic realities, defined on the individual level prior to the perception of the other, correspond to the experience of animal life, we see that the 'transcendental constitution' (as the very disposition of constitutive analyses in the manuscript of *Ideas* II demonstrates) takes up again, only on the abstract plane of the lived, the real changing of *matter* to *life* and of *life* to *spirit*, understood as *social* existence. It is true that at each stage an original structure arises: "to consider these founded unities with no prejudices, if we bring them back by the phenomenological method to their sources, they are precisely grounded and of a *new type*; the new element that is constituted with them can never be reduced (as the intuition of essences teaches us) to the simple sum of other realities." [3] But the phenomenological relation of foundation implies precisely the intelligibility of the passage from the *founding* level to the *founded* level. It is not a question of a 'reduction' of the superior to the inferior, but of a *dialectical* movement in which the relations that develop at the interior of a given form, move in a manner necessary to the constitution of a radically new form. From that time, *materiality* (*Dinglichkeit*) is not a simple substrate indifferent to the significations which it bears. It defines the originative resting place from which the movement engenders more elevated modes of being in the specificity of their meanings, the real *infrastructure* which founds the ideal *superstructures* in their historical emergence and in their truth value.

Thus, concrete phenomenological analyses can grasp all their meaning and be developed fully solely on the horizon of dialectical materialism. It goes without saying that we are obliged under these conditions to reject not only the totality of the Husserlian doctrine but also the method itself to the extent that it has become ossified in abstract formulas. In addition, the concept 'transcendental' was superfluous from the outset, since it maintains a strict identity of content between 'pure consciousness' and natural consciousness. Be that as it may, theory has meaning only in terms of practice, and the practical demands of working out a description require the overturning of the theory of transcendental idealism. What semblance of reason would there be in being obstinate in limiting to the purely lived the study of kinesthetic and sensible configurations, when one is concerned, obviously, with elementary sensori-motor formations that are covered over at an adult age with an enormous cultural acquisition and are uncovered in the pure stage in only an animal or an infant? – Likewise, the historico-intentional analyses of Husserl's last period, so rich in suggestiveness, are extraordinarily deceptive in their uneven character and their lack of real content. But it is certainly not a question of weaknesses of a personal order, since we have seen the unequaled mastery of Husserl in this matter. Rather, the very horizon of phenomenology turns the gaze of the phenomenologist away from the real data that, on the other hand, define the true content of his reflection. More precisely, the real data appear only under their negative aspect in view of the fact that their signification has been suppressed. But the actual movement of description returns ineluctably to this material reality, which is revealed consistently as the ultimate resting place of constituted formations. Thus we find ourselves confronted with an intolerable contradiction that obliges us to pass to the point of view of objectivity in freeing ourselves from the theoretical concepts of phenomenology in the name of the technical necessities of the descriptive method.

But then, once more the ghost of psychologism rears its head. How would it be possible to justify within the framework of material nature, (i.e., animal and social) the *truth* to which the intentions of consciousness lay claim – truth that one can dispute in particular cases but that no-one would know how to deny in principle without at the same time denying himself? It is I, only a single being, an object among other objects, who carry the world in the spiritual interiority of my lived acts, and the world is *in me* in the very same operation by which I perceive myself *in it*. It is here that the existentialistic tempation is presented which seems to offer a convenient means of ratifying all the real data of existence in the world, all achieved by maintaining a metaphysical opposition between man and nature. *Being-in-the-world*, Heidegger assures us, is not an objective circumstance which would impose itself because of the reality of things, but rather an ontological structure that belongs in its own right to the existing human being: man exists not because he is in the world and not by reason of his position in the world, but his position in the world is possible precisely and only because he exists as man, and by reason of his human essence. So, the fact that man is only one being existing among many others presents no difficulty, since man carries the world in the *project* of his being, and it is by this very project that he is constituted as a being-in-the-world. To put it another way, the actual problem posed by the structure of intentionality – to know that the world is ideally in my consciousness when I am really in it – finds its immediate solution in the magic of language which by a simple reordering of words, transforms the expression '*the world in which I am*' into a moment of myself, inasmuch as I am precisely *being-in-the-world*. The common man would be satisfied with saying: "Man is in the world"; the existentialist philosopher exorcises such naiveté and moves on to the level of 'existential ontology' in assuring as the '*man is being-in-the-world*'. All that is left is to dissect the expression in some way or other, or to detach the 'being within' (*In-Sein*) and to present it as an original moment, and everyone sees without difficulty that the 'world' is nothing more than an element of 'being-in-the-*world*', and thus of man.

Mystification is a common procedure for philosophers. At least the classical tradition, of which phenomenology represents the ultimate form, has had the elementary good sense of reproducing on the symbolic level of ideas or consciousness the real operations by which man has transformed nature and rendered it assimilable to his thought. One such transposition permitted an at least formal justification of existence in this world by science and reason, in which the dignity of human labor is reflected. With existentialism all rationality is abolished on behalf of a 'project' which claims to appropriate the reality of things from now on without providing itself with any foundation that legitimates its claims. More precisely, under the pretext of reuniting the concrete data of 'existence', it is the very *absence* of *justification* that is erected resolutely as the *supreme justification*, within the arbitrary absolute of a '*liberty-unto-death*'. All the values acquired by the long effort of the humanist tradition are discovered to be suddenly denuded of real foundation and are maintained now, only by clinging to the '*resolute decision*' of defending *even to death* what one can no longer justify in terms of reason and truth.

It would be of little use for us to delay in considering the innumerable inferior imitators that have proliferated on the Heideggerian model. The great problem of our time, in which is expressed the feeling that has become unanimous, that the ideal subject of traditional religious or philosophical thought be identified rigorously with the real man in this world, has for a long time found its solution in the Marxist dialectic which defines the only valid process for a constitution of lived significations on the foundation of material reality. The notion of *production* takes into full account the enigma of consciousness inasmuch as the object that is worked on takes its meaning for man as *human product*. The realizing of meaning is precisely nothing but the symbolic transposition of material operations of production into a system of intentional operations in which the subject appropriates the object ideally, in *reproducing* it in his own consciousness. Such is the true reason for which I myself, who am in the world, 'constitute' the world in the interiority of my lived acts. And the *truth* of any constitution of this sort obviously is measured by the actual power of the mode of production from which it takes its model. But the philosopher remains ignorant of these origins. Inasmuch as he is a member of an exploiting class, he does not have experience of the real labor of exploited classes, which gives things their human meaning. More precisely, he perceives this labor only under its ideal form, in the act of *commanding*, and asks himself with astonishment how these 'intentional significations' were able to be imposed on the real world. The reflection involved in self-consciousness, evidently, can only confirm these intentions themselves in their lived purity and place them 'outside the world' as pure constituting syntheses in the *'liberty of spirit'*.

Thus, the social relations of production and the division of society into classes hinders the ruling classes from giving an account of the real foundation of ideal values, by which they claim to demonstrate their human quality and to justify their domination. Exploiting the labor of the oppressed classes, they perceive the produced object in its human meaning, but this meaning appears to them only in its pure ideality, negated of all material reality, since they certainly mean to take no material part in its production. As a member of a dominating class I accede to the *truth* of being only in *denying* being that is effectively real, the real labor of the oppressed classes which I 'go beyond' in the intentions of my consciousness only from the very fact that I appropriate its product. *The form of oppression is the key to the mystery of transcendence*, and the hatred of *naturalism* does nothing but express the *natural* repugnance of the ruling classes to recognizing in the labor that they exploit the true source of meanings to which they lay claim.

The difficulty in understanding the real genesis of ideal significations found itself once again reinforced by the abstract manner in which materialism was elaborated in bourgeois thought in the course of its struggle against feudal power. During its revolutionary ascendency, when it represented the general interests of human society, the bourgeoisie was already an exploiting class, even though its position as oppressed did not yet allow it to organize itself apart from the laboring masses. Also, the materiality of productive labor, whose memory it still kept alive for use in opposing the spiritualism of feudal exploitation, manifested itself only on the abstract horizon of its own mode of exploitation. In the hands of rising capitalism, the concrete products of the earth and the workshop were reduced to the pure abstract universality of their exchange value, as simple calculable moments in the movement of money. From that time the creative power of material labor could not reveal itself in its real effective process as the very dialectic wherein all sense of truth is engendered, but could do so only under the abstract form of a pure mechanism in which it became available for new relations of production. Moreover, when the bourgeoisie, having arrived at the decisive phase of its struggle for power, finally ceased to conceal its naturalism under the protective veil of natural theology, in order to be able to affirm the absolute value of human labor through the concept of matter, it maintained itself necessarily within the limits of mechanistic abstraction, viewing this same labor only as it had exploited it on the level of the abstract rationality of economic calculation. – Very clearly, defending human interests solely in terms of its own mode of exploitation, once having achieved political domination, it had no other concern than to ally itself with the previous ruling class in order to oppose the new humanism that was rising up among the proletariat. This very same materialism that had carried it to power now served it as a scapegoat, in order to depreciate the effective reality of productive labor. The interpretation of matter as pure mechanism permitted a facile condemnation that systematically confused the creative materiality of the laboring masses with the *sordid* materiality of capitalistic exploitation. The critique of 'psychologism' obstinately set itself against a phantom, which reflected within the consciousness of the oppressor that human reality which he divested of all human meaning.

Nevertheless, in taking as a pretext the 'defense of the spirit' in order to drive back the effective movement of human progress, bourgeois thought cut itself off from the true source of its own spiritual values and for this very reason ended in its own internal dissolution. If Husserl still remained within the tradition of idealistic rationalism, showing signs of the late flowering of the German bourgeoisie and its final radical whims, his evolution bore witness to nothing less than an increasing uneasiness with regard to the *real foundation* of meanings apprehended in consciousness. From the enjoyment of eternity to the intuition of essences to the anguished problematic of the *Crisis of the European Sciences*, interpreted as the crisis of Western man, the feeling that traditional values had become bankrupt grew stronger every day, and the famous rallying cry, 'Return to the things themselves', took on more and more openly the meaning of a return to the sensible realities of the *life-world*. But Husserl's class position did not allow him to go back to the social relations of production that defined the real content of sensible life; and the 'transcendental constitution', reduced to seeking its ultimate foundation in the pure sensorial datum, ended paradoxically in a complete scepticism. In the meantine, Heidegger had deliberately renounced the classical demands for rationality in order to limit himself to the pure 'transcendence' of the project of being, in which, by virtue of philosophers, the *absence of reason* is transmuted into the *supreme reason* of existence. In the decomposition of bourgeois society that had been caused by the ruthlessness of imperialist monopoly, the motto *'liberty-unto-death'* was offered to the ruined *petite-bourgeoisie* as the final justification of their position as *petite-bourgeoisie*. With the passage from phenomenology to existentialism, the disdainful theme of the critique of psychologism, the denial of human subjectivity in the name of objectivity and the universality of Truth – revealed its own inconsistency by giving way to the pure negation of the real conditions of existence in the pure subjectivity of 'resolute decision', where individual arbitrariness systematically was erected as the ultimate foundation of all true value.

Abstract materialism became all the easier to refute as it was reduced, in the final analysis, to a shameful idealism, with pure mechanism being identified, as well, with pure thought. Such discussions, evidently, did not in any way affect dialectical materialism which takes its meaning from the proletarian experience of creative labor. In the real process of production *man is homogeneous with matter*, and it is in that material relation itself that the original relation of consciousness to the object that it perceives is constituted as 'constituted meaning'. But in the past the laboring masses had not been able to raise their sights to an ideology that properly belonged to them and exactly expressed the structure of their productive activity. In fact, the weak level of productive forces involved a multitude of conflicts in which destruction appeared, and whose transposition on the symbolic plane of consciousness covered over the objects that were originally economic with a cloud of spirituality that defined the meaning of *sacrifice*: From that time, effectively constituted significations in productive labor became alienated in the *transcendence* of an ideal negation of that same reality; and thus, a certain number of people took the opportunity to appropriate the means of production for themselves. Such a mystification in which *appropriation* was accomplished in the form of an *expropriation*, resulted from the very movement of production and, in fact, included the totality of producers. More precisely, the perspectives of the exploited still envisioned merely an eventual passage to the position of exploiter, and, while the struggle continued between various types of exploiters, the oppressed masses could only give their support to those who seemed, at first glance, the most easily accessible. Thus, without going explicitly into all the subtleties of ideological conflicts and all the details regarding the interests at stake, one can say that their conception of the world was indistinguishable from that of the exploiting classes and altogether contributed to perpetuation of the general form of exploitation. *Religion* was the expression of this unanimous pretension to privileged positions, idealized in a supreme transcendence in which each retained the hope of some day profiting from the regime of oppression. It is only with the development of mechanization and of major industries that a new class of exploited appeared which had the experience in its daily practice of new relations of production in which the common exploitation of nature by human society replaced the exploitation of man by man. In this new mode of existence that necessarily has been begotten in the very womb of capitalistic society as the form of its suppression, the perceiving subject is no longer the real or virtual exploiter who expropriates the producer by denying the reality of his product, but rather the producer himself, who defends the materiality of his production against mystifying idealizations. Since the new productive forces had worked out the conditions for an appropriation of labor by the workers themselves, the proletariat (freed by the very brutality of capitalistic exploitation from all hope of arriving in their turn at a position of exploiter) are able to see in the spiritual values of the previous society nothing other than bourgeois prejudices which dissimulate the sordid materialism of bourgeois practice. To be sure, the proletarian movement has not been limited to the factory workers who constitute its authentic core and its permanent foundation: it has consistently absorbed increasing strata of previous exploiting classes that are crushed more and more by the mechanism of capital and whose adherence to communism realized the dialectic of bourgeois society in the *truth* of its becoming. In fact, we are reminded each day that communism defends the true content of traditional values, by means of which its proponents can still retain some understanding of the permanent dissolution imposed on them by the bourgeois society. Ideal aspirations, reduced by capitalistic exploitation to simple forms of hypocrisy that betray themselves through the evidence of their futility, recapture a human meaning by being integrated with the constructive tasks of the proletarian revolution which, in the face of the profundity of the bourgeois decomposition, takes charge of the general interests of mankind and pursues the construction of socialism in the very course of the anti-imperialistic conflict. The *truth* of dialectical materialism is demonstrated in the *real historical dialectic* in which, under the pressure of continual proletarization, the former ruling classes become progressively more aware that their ideals are the simple results of their material conditions of existence. From this it follows that materiality is the authentic origin of all meaning and value.

Thus it is found that traditional problems are completely resolved within the framework of Marxism, and there is no longer any reason to hesitate in drawing from *technical* difficulties that we have encountered in phenomenology their *philosophical* consequences, in the proper sense of the adjective. Once the real content of lived structures has been recognized, a technical solution would have to be limited to the 'completion' of intentional descriptions by means of objective analyses. But in their sense of reality, the real data are totally incompatible with the phenomenological absolute of lived intentions, and such a *mosaic* would unfailingly have entrapped us in inextricable contradictions that would have revived, along with the terror of psychologism and the myths of transcendence, practically insurmountable obstacles to positive research. The principal merit of phenomenology was its definitive destruction of formalism within the very horizon of idealism and its placing of all problems of value on the level of the concrete. But the concrete can be described correctly only in the actual movement of its material determinations, a fact that implies a total liberation and the passage to a radically new horizon. Nevertheless, we believed that it would be useful to present, in the first part of this work, studies which are strictly phenomenological and mostly outdated. These demonstrate, and do so better than any systematic critique, the internal necessity of the ground covered. From the eternity of essences to lived subjectivity, from the singular Ego to the universal genesis, the evolution of Husserlian thought has borne testimony to the constant aspiration of idealism towards that real content whose authentic conception can be defined solely in terms of dialectical materialism. In Marxism, bourgeois philosophy finds the form of its suppression: But suppression includes the very movement of what it suppresses, insofar as it *realizes* it in *suppressing* it.

T. D. T.

PART ONE

THE PHENOMENOLOGICAL METHOD AND ITS ACTUAL REAL CONTENT

"Mankind poses only problems that it can resolve."

(Karl Marx)

CHAPTER ONE

THE INTUITION OF ESSENCES

1. THE TECHNIQUE OF VARIATION

Phenomenology began as an ontology. It was a question of going beyond psychologistic interpretations which transformed the real into a collection of states of consciousness so as to return to *the things themselves* and to rediscover the *meaning of being* in the fullness of its truth. The first effort of Husserlian analysis set its sights precisely on the very notion of *truth* which had been obscured by the reigning empiricism. It aimed at restoring the concept of *being* in all its dignity, and the enthusiasm that welcomed the first volume of the *Logische Untersuchungen* demonstrated that the *authenticity* of human existence itself was also at stake.

While contemporary logicians viewed science as a product of subjective consciousness and remanded logical laws to psychology, Husserl, with unequaled precision, restored the very meaning of knowledge as knowledge of *truth*. When I affirm a proposition as true, it is clear that I take it in its ideal objectivity as valid for everyone and for all times. Therefore, logical conditions, without which there would be no possibility of any truth, could hardly depend on psychic states of the real subject. They define the *essence* of truth such as it is *in-itself* and such as the skeptic himself would necessarily recognize, inasmuch as he upholds a theory and affirms it as true. And, in fact, when the logician declares that "of two contradictory propositions, one is necessarily false," he does not base himself upon the observation of acts of consciousness, but rather on an intellectual intuition in which non-contradiction imposes itself on him as belonging to the very *essence* of any valid statement. Thus the critique of psychologism, so brilliantly enunciated in the first volume of the *Logische Untersuchungen*, succeeded in reestablishing the validity of the notion of essence by founding it on an 'evidence' of a special type (*Einsicht in das Wesen, Ideation*).[1] From then, the very movement of the demonstration implied the possibility of an indefinite generalization: since the radical failing of psychologism was to dissolve the object into a succession of states of consciousness, its absurdity would be demonstrated for every mode of being. Each existing thing would have to be defined, at the outset, in terms of its own essence without which it would have no meaning to its listeners. However, this consequence was not yet made explicit in the first volume which we have just cited. It lacked a method of thematization, which alone could give an actual signification to the theory. The precise definition of the intuition of essences in the second volume was to allow for the elaboration of a doctrine of being, such as was imposed independently of the contingent circumstances of its realization, inasmuch as it found in this doctrine the *a priori* possibility and, at the same time, the truth of the subjective acts of the consciousness which grasps them.

But, strange as it may seem, the technique of the *Wesensschau* appeared at times in a relation familiar to the empiricists: the relation of the whole to its 'non-separable' parts, such as color and extension in the visual object.[2] It is important to remember that herein lies the basis for Berkeley's argument: there cannot be an 'abstract' idea of color or of extension, precisely because each of these cannot be posited 'apart'. The originality of Husserl consisted in drawing out the *meaning* of this remark: if it is impossible to have a color without a surface, the result is that it belongs to the *essence* of color to appear only on a surface. Husserl saw within this, which was the basis for Berkeley's belief that he had refuted the possibility of escaping from the perceptible or sensible, the precise means of thematizing pure idealities. Each time that we establish an 'inseparability' of the kind just described, we can make use of it to define an 'essence.' No matter how much I would try to imagine all possible colors under whatever forms I choose, I could never suppress the surface, since color would then disappear at that very same instant. The *consciousness of impossibility* defines a *condition of possibility* – an *a priori* law.

We might wonder whether what we take for an essence is, in fact, only a property of the imagination. The impossibility in question could belong to empirical circumstances. In fact, the author is less concerned with the intuition's being realized than with the possibility of *the thing itself*. Thus it is clear that concrete objects are no longer perceived 'apart': they always stand out against a background, the nature of which is often imposed by force of habit. But all the same, they *could* exist apart – it is *imaginable*, that is to say, *thinkable*. In the case of color, we run up against an absolute impossibility that resides in the very nature of things. It is not a question of a psychological condition for intuitive representation (e.g., that I cannot describe a head to myself without a body appearing at the very same time, more or less clearly, to complete it). Rather, it is an ontological condition of being itself. There cannot be color without extension, because this is *unthinkable.*[3]

This text should not present any mystery to the reader familiar with the notion of intentionality. Whereas the empiricists, viewing states of consciousness as things, could define the non-separability of color only in terms of the impossibility of realizing a separate sensation (in other words, the *impossibility of a consciousness*), Husserlian analysis demonstrates that it is a question of a *consciousness of an impossibility*. It is the nature of the object itself which is perceived and thus known in such an experience. The argumentation of empiricists is reversed by a *thematization*,[4] in which the subject is revealed as the bearer of the meaning of being. The intervention of intentional analysis gives rise to a dialectical reversal that allows us to reach an objectivity by subjective description.

Thus we see the outline of a technique that permits the systematic study of the realm of essences. The essence as the condition of possibility will be revealed in a consciousness of impossibility. Whatever one cannot eliminate without at the same time destroying the object itself is an ontological law of its being, belonging to its essence.

The procedure by which we methodically create this consciousness in ourselves is called *variation*.[5] We start with any object whatsoever as 'model'; we 'vary' it in a totally arbitrary manner. It then becomes apparent that freedom cannot be absolute, that there are conditions without which these 'variants' could no longer be variants of this model, as 'examples' of the same kind. This 'invariant' factor, identified throughout all differences (*Deckung im Widerstreit*) defines precisely the *essence* of objects of the same species without which they would be unimaginable, i.e., unthinkable.

We should note that variation is not, strictly speaking, a game of the imagination. It is not a matter of disengaging the common element from all cases that we have actually imagined, since, in fact, their number is necessarily limited and the *invariant* has to impose itself on all *possible* cases. What is important is 'the form of the arbitrary' (*Beliebigkeitsgestalt*)[6] in which the movement unfolds itself and of which we become immediately aware in the lived feeling of 'I can'. Possibility and impossibility are put to the proof of factual evidence.

In this manner the original intuition, the 'perception' of essences, is realized. The *eidos* 'itself' is grasped as the *being* of the object, such as it *is* necessarily inasmuch as it is an object of such a nature. Thus the 'thing',[7] the object of sensible perception, is a spatio-temporal ensemble, provided with 'secondary' qualities, posited as a substance and a causal unity. Such a definition constitutes its *essence*, inasmuch as it would no longer be a 'thing' if it lacked one of these elements. We can put this to the test through variation. The essence in this instance, is revealed 'in person' as invariant. It is here 'given' to us, not, to be sure, as a 'thing' but precisely as an 'essence' in the mode of evidence that is proper to it. The notion of an eidetic intuition is nothing like a metaphysical hypothesis: it imposes itself in its very actualization. The existence of the *eidos* is demonstrated by its very presence which we can obtain through variation.

2. PURE IDEALITIES AND EMPIRICAL IDEALITIES

The extension of the notion of intuition to 'universal objects' generally provokes a hostile reaction: is this not realizing an idea or making it into a thing, a datum, which one must receive passively from the outside?[1] This objection rests simply on a misunderstanding. The intution, as phenomenology conceives it, is in itself neither active nor passive. It is defined in its actual form solely in relation to the kind of being which is in question. The intuition of idealities will be properly revealed as *active* and even *creative* (*Erzeugung, Schöpfung*).[2] The notion of operation does not contradict that of intuition, since the idea 'itself' can be given only in an operation. The intuition is nothing more than the very act of knowledge, insofar as knowledge is seized from being, in opposition to the empty intentionalities of 'simple discourse'. Inasmuch as the essence can be the logical subject of legitimate predications it must be able to be 'given', since what is true of it is true of 'itself' and refers necessarily to an 'original intuition' in which it 'itself' is grasped.

The doctrine of the plurality of modes of intuition and, correlatively, of modes of existence, corresponds to the very structure of the phenomenological evidence. But the eidetic method poses another problem of a much more serious nature. The *eidos* is a pure ideality,[3] radically independent of every actual perception of real individuals. The *Wesensschau* defines a system of *a priori* laws which imposes itself on factual science inasmuch as they must always conform to the essence of their object. We can then ask what can warrant their intrusion on 'experience', in the ordinary sense of the term.

Evidently, there could be no question of confusing variation with the common method of abstraction and generalization. We have pointed out already that the invariant is not obtained, strictly speaking, by means of a comparison, since it must be common to all *possible* variations, and one can in fact imagine only a small number of them. It is revealed in a consciousness of impossibility that is brought about by the presence of the essence *itself*. Empirical idealities, objects of ordinary concepts, are connected to their actual realizations. Their extension which is *de jure* infinite, is *de facto* limited by the individual cases upon which they have been established. Should experience give rise to new data, we will be forced to revise them. Any such danger could not threaten the *eidoes* which does not depend in any way on examples that we have been able to give of it: they are, by definition, what is possible solely because of it.

But another objection is offered. Variation is not confused with an abstraction based on an empirical comparison. But is it anything other than a purely *logical* analysis of concepts as such? The *invariant* is that without which the variants would not longer be variants of the same *model*, objects of the same *genus*. Thus, it is only a question of defining that genus in itself. Its 'essence' is only the content of its concept, which, inasmuch as *already constituted*, is posited, evidently, in its intelligible being as independent of 'experience' in the ordinary sense of this term. But was not that very content originally disengaged from empirical data? If I attempt to obtain the essence of a swan by variation, I will find whiteness among its eidetic attributes. A swan, were it not white, would not be a swan. But that was true only until the discovery of black swans. Variation is a convenient process of explicating the meaning of our concepts, and specifying what we mean when we speak of such and such an object. But the concept refers back to its *origin*. It is always permissible to analyze a notion that we already possess. But the whole question is to know how, in fact, it has been constituted.

The answer of the author is perfectly clear.[4] We do not concern ourselves in the least with the empirical genesis of our representations. Every concept can be taken in its own content as pure possibility. There is, evidently, an infinite plurality of essences of which only some are realized. But each one exists in itself, in its ideal being, independently of the conditions for its insertion in the world: accordingly, it can always be defined as such. The *eidos* is independent of experience – in the ordinary sense – because we do not refer in variation to any experience, real or possible. The examples are taken as pure possibilities, solely with an intelligible content, and without any relation to any situation whatever in the real world. The essence defined in variation is presented as absolute, in the mode of existence that is proper to it, namely *pure possibility*.

So, we no longer have to raise the problem of the distinction between the empirical and eidetic. Every concept taken by itself designates an essence. There will be an essence of tree, an essence of red, an essence of centaur. Here, we rediscover a notion that is well known in the philosophical tradition. But difficulties, equally as classical, are going to reappear immediately. We seem to eliminate the old problem of the relations of essence and existence by taking refuge in the realm of pure possibility. However, can we remain there indefinitely? In fact, the authentic motivation of eidetic studies would demand a return to the real.

3. THE TRUE SIGNIFICANCE OF THE NOTION OF ESSENCE

The reduction of the *eidos* to the pure possible, protected it *a priori* from every encroachment of experience. In this way, the 'Friends of the Forms' had shielded themselves from the attacks of the 'Children of the Earth' in the celebrated Gigantomachia which Plato tells of in the *Sophist*. But to set the intelligible 'apart' is to render it useless. In fact, essence in its original notion did not concern itself with the *pure possible*, but rather with the *possibility of the real*. It is placed above the actual object only in order to define the *meaning of its objectivity*.

Already from the outset, we have pointed out the ontological signification of the theory. The first volume of the *Logische Untersuchungen* opposed to skepticism the necessity of respecting those conditions without which no theory would be possible, not even the skeptical theory itself. These conditions constitute the very *essence* of truth, that which assures that the truth *is* truth. Very evidently, it is not a question of a 'possible' truth in the sense in which it would be posited absolutely apart from actual truths. The essence of truth defines the conditions of all *possible truth*, understanding by this last expression not some ideal being that is 'simply possible', but *truth itself inasmuch as it is possible*. This possibility is its *very being*, that which it *is* authentically.

The intuition of essences in its original form deal directly with the very being of the things, *the being of the existent*. The eidetic investigations do not aim at the simple analysis of concepts but the constitution of a *universal ontology*. Before studying the facts, it is expedient to define the essence which constitutes their being. Physics is possible only from the moment that one already knows what a physical fact *is* as physical. And, actually, the science of nature has existed as a science only from the day that Galileo discovered that the *being* of the physical consisted in its *being measured*. The same condition is placed on all positive research in general, inasmuch as the conceptualization of its facts must conform to the meaning of their being. Thus, we shall never have a valid psychology as long as we have not yet defined the *eidos* of the psychic, that which makes the psychic be psychic, that which must be the object of our attention if we want to take hold of the psychic as such. Each kind of existent, inasmuch as it possesses an original being, constitutes a particualr domain. Before engaging in experimental research, it is expedient, then, to state precisely the *region* in which it will be pursued. It is clear that *being* does not have the same meaning for the physicist as for the psychologist. What *exists* for the former is a certain spatio-temporal unity, posited as an *in-itself*, in causal relation with others that are *in-themselves*. The being of the psychic, on the contrary, is merged with its appearance: it is a monadic unity in the flow of the lived. The eidetic phenomenologist assigns himself the task of describing diverse *regions of being* in their own structure, in a manner that makes precise the *pure part* of the positive sciences which studies them in experience. The eidetic laws determine the *conditions for the possibility* of empirical knowledge – different for each of its domains. They are independent of the real, not because they are outside it, but because they precede it and give it the *meaning of its being.*[1]

Such is the original signification of the notion of essence. A more radical elucidation would have led us to carry out a return to a constituting 'I'. The essence, as meaning of being, obviously refers to the intentionalities of the transcendental consciousness. But the eidetic method studies it on the level of object – a convenient attitude and necessary to prepare the field for more profound studies. Conceptual analysis of the essence of diverse regions of being outlines for us the framework for the study of the acts which *constitute* them.[2]

But the method surpassed its object. Variation did not apply solely to concepts that were strictly ontological, but also to all concepts in general. Thus, there will be an essence of house, an essence of centaur, as well as an essence of truth. The thematization of the *eidos* as object left only one means of preserving it from empiricism – to define it as a pure possible. But this was to take away its very *raison d'être*, its relation to the real. The essence is, authentically, the constitutive meaning of being. In variation, it is merged with 'simple significations'.[3]

4. DIFFICULTIES WITH THE OBJECTIVISM OF ESSENCES. THE RETURN TO THE SUBJECT

We will not insist on the multiple problems which are raised, on the philosophical plane, by the description of essences. Its true role is simply preparatory. The convenience of the method was able to create enthusiasm among its first disciples. But the *eidos*, despite appearances, is still in the world; and phenomenology, properly speaking, begins only with the reduction.

But if it would be out of the question for us to work out here a detailed refutation of the wordly doctrine of essence[1] (such as is elaborated at the heart of the natural attitudes), it is nevertheless expedient to indicate, at least, the character of its difficulties, necessary to the extent that they are derived from the 'essence' of logicist objectivism – in the authentic sense of essence. As we have seen already, treating the *eidos* as an object involves condemning oneself to confusing the *possible inasmuch as it founds the real* with the *pure possible* defined by the exclusion of every relation with the real. Thus, 'mathematical existence' is characterized precisely by the elimination in principle of every *problem of existence*.

This lapse of the ontological into the ontic is translated in still another manner. Just as every signification, considered in itself and abstracted from every relation to its object, constitutes an essence, so the eidetic domain becomes inordinately extended. Every existent, taken in its pure intelligible content, will have its singular essence. As far as logic is concerned, it will not be sufficient to study the meaning of the notion of theory in general: we have to construct all the forms of possible theories.[2] The eidetic of space will not be simply the philosophical study of the concept; it will encompass the whole of geometry. Both material and formal mathematics are qualified as sciences of essence: this assimilation, because of the confusions that it entailed, could only discredit the phenomenological method.

These faults belong to objective idealism as such. The realization of essences is the first step for philosophy, and the difficulties that it raised are nothing but the exigency of a dialectical surpassing of them. The discovery of that fundamental philosophical truth, that the existent presupposes the meaning of its being, is inevitably interpreted at the outset by positing this meaning as an existent of a special category. Thus, philosophy began by affirming that beautiful things exist only through Beauty-in-itself. The Idea was objectified by being placed in a world 'apart'. From that time, authentic essences, defined by their ideal being, as pure significations, were confused with the content of ordinary concepts, and philosophy retained its meaning only by means of the feeling that the philosopher had for his vocation.

If things exist solely through their participation in the idea, said old Parmenides to young Socrates, what must we think of objects such as hair, mud and dirt? – Not at all, said Socrates. In these cases, the things are just the things we see; it would surely be too absurd to suppose that they have a form. All the same, I have sometimes been troubled by a doubt whether what is true in one case may not be true in all. Then, when I have reached that point, I am driven to retreat, for fear of tumbling into a bottomless pit of nonsense. Anyhow, I get back to the things which we were just now speaking of as having forms, the Beautiful and the Good and other things of the same kind, and occupy my time with thinking about them.[3]

This is the philosophical movement that Husserl reenacted at the beginning of the twentieth century. The astonishing freshness of his undertaking can be called a *new creation*. His originality was manifested in the profound expertise shown in his method. Thanks to variation, concepts could be analyzed with a precision until then unknown. They undoubtedly lacked a philosophical foundation, but the method remained, nevertheless, perfect in itself – a perfection that was due to the authenticity with which the author experienced the presence of the Idea.

The first thematization could take place only on the level of the object. It met with difficulties that motivated its displacement. That the essence determines the *being of the existent* can find its true meaning only in a return to the *subject*. That, if we take the *eidos* simply as object, its independence in relation to experience can be guaranteed only by the elimination of all reference to the real; but by the same token we have excluded the very goal of our endeavor, to know the constitution of the world through the unveiling of the meaning of its being. In fact, *meaning* refers to the *act*. The *being of the world*, that which it *is* authentically, is what we aim at when we speak of it. The task of a universal ontology can be realized solely through the explication of the *intentions of the constituting consciousness*.

The passage to subjective idealism does not, in fact, posit anything really new. Eidetic analysis was an intentional analysis which was not aware of itself. The invariant is imposed upon us, because, if we were to suppress it, the object would no longer have the same meaning: it would no longer be the same *for us*. The consciousness of impossibility was nothing more than the translation of a consciousness of *signification*. Yet, signification is what a thing is for a subject, its *being-for-me*. Thus, the return to the lived, was only a clarification. The analysis of the essence of being discovers its true meaning by referring to the Ego, since the essence is only the meaning that being presents through it. But making this clear is realised by an absolute reversal. While in the *Logische Untersuchungen* being is posited as an *in-itself*, it now reveals itself only as being *for us*. The second phase of Husserlian thought begins with a *metabasis eis allo genos*: the transcendental reduction.

CHAPTER TWO

THE THEMATIZATION OF CONCRETE CONSCIOUSNESS

5. THE RETURN TO LIVED EXPERIENCE IN THE *LOGISCHE UNTERSUCHUNGEN*

The first volume of the *Logische Untersuchungen*, published in 1900, had approached the problems of logic from a strictly objective point of view. The colossal project of a *Mannigfaltigkeitslehre* in which all forms of possible theories would be deduced in a systematic manner, returned to the old dream of a *Mathesis Universalis*. The mathematical education of the author,[1] who once had been an assistant to Weierstrass, seemed to destine him to a brilliant career as a logician. The publication of the second volume, in 1901, was a surprise. The problem of the theory of knowledge, discreetly announced in the first volume, [2] was developed by a return to subjectivity.

However, the preface of the work marked rather clearly his break with the psychologistic period. *The Philosophy of Arithmetic*, published in 1891, had presented as its task the founding of mathematics on psychology,[3] according to the dominant tradition of the period. But if the method seemed to give an account of the genesis of representations, it was revealed as incompatible with the objectivity of science. Consequently, a radically new foundation had to be found.[4] Thus, the second volume of the *Logische Untersuchungen* could not be explained by a return to psychologistic methods. The apparent similarity in the formulas serves to conceal the originality of Hussarl's inspiration.

A remarkable text of the unpublished part of the *Krisis* provides us with enlightenment here. The author tells how he discovered, during the writing of the *Logische Untersuchungen* around 1898, the universal correlation between subject and object. Every existent, no matter to what domain of being it belongs, is an index for a system of the lived in which it is 'given', in accordance with *a priori* laws. This revelation, declares Husserl, "affected me so deeply that my whole subsequent life-work has been dominated by the task of systematically elaborating on this *a priori* of correlation."[5]

So we see the radical newness of the original intuition which inspired phenomenology. The notion of intentionality in Brentano still designated only a characteristic of 'psychic phenomena' and permitted the subsistence of a reality existing in itself and beyond its reach. From the period of the *Logische Untersuchungen*, Husserl reaches the central point where subject and object appear as inseparable. The hypothesis of an *in-itself* becomes as absurd as that of a consciousness that would not perceive the world *itself*. At the same time that psychologism is rejected, abstract realism is rejected for a philosophy which sets itself the task of describing the manner in which being is for the self and the manner that the self knows being. At issue is not the relation between two realities which could exist apart, but of an *essential* relation, defined by *a priori* laws, without which neither consciousness nor the world would be conceivable. The subject to which we returned was from then and thereafter a constituting subject:

> The further course of the explanations in this text will show how, when human subjectivity was brought into the problems of correlation, a radical transformation of the meaning of these problems became necessary which finally led to the phenomenological reduction to an absolute, transcendental subjectivity.[6]

Thus, the consciousness thematized in the second volume of the *Logische Untersuchungen* was a 'neutral' consciousness,[7] which will not be defined as transcendental until the problem of psychologism is posed. At the beginning of the investigation, the subject was constituting *in-itself* without being yet *for-itself*. It is from this point of view that it is appropriate to comment on the motivation that the author presents to us in his introduction.

The return to lived experience is not occasioned by any curiosity of the psychologist but by the demands of the logician himself. At stake is making specific the meaning of logical objects, and to accomplish that, abandoning representations which are more or less obscure in order to return *to the self-same thing*. But the *thing*, that which we intend when we speak of logical objects, is what is present to us when they are given to us without the *evidence*. When presented with the formula of a logical law, one should not be satisfied with a simple verbal understanding. The concepts have to be rendered present in fully carrying out the operation of 'idealizing abstraction' which makes them stand forth before consciousness, in order to verify that this 'given' is indeed what it is said to be. Many times have logical laws been interpreted as psychological laws on the basis of a merely 'symbolic' comprehension. The return to evidence allows for elimination of these confusions. Thanks to the determination of the conditions for such evidence, we shall use concepts which will make precise the meaning of the object in question: such is the aim of *phenomenological* analysis.[8]

Based on this fundamental text, we can explicate the original significations of the return to the subject. Logic is the science of concepts, propositions, relations, etc. These 'logical objects', as all objects in general, are posited outside of the knowledge which intends them. Their properties are defined by the propositions of logic. Thus the law that "of two contradictory propositions, one alone is true" expresses a property of a certain category of logical objects: propositions. Reflection can then take two opposite directions. On the level of the natural attitude, the logician, as every scientist in general, will try to discover as many laws as possible and will try to codify them in a coherent system. The ideal achievement of this work would give us the 'general theory of formal domains' (*Mannigfaltigkeitslehre*) where all possible statements, taken in their *form*, would be deduced in a rigorous manner. All imaginable theories would be constructed, thus, *a priori*, and there would be nothing more left than choosing the one theory that would conform best with the data of experience. Realizing the ideal of *Mathesis Universalis*, the *Mannigfaltigkeitslehre* would be, on the level of the object, the theory of theories, the science of sciences.

That is the conclusion of the first volume of the *Logische Untersuchungen*. But the need for intelligibility can take a different course. The deductive science of logical objects begins with fundamental concepts which the logician accepts without criticism: the direction of his interest extends only to their use in the system that he constructs. From this purely objective point of view, he possesses a sufficient knowledge of them. The scope of logic is precisely only knowledge itself in its realization. The logician speaks of concepts, propositions, relations, truths. But what is a concept, and what is the signification of conceptual knowledge? What is truth, and how can an object (seeing that it is, as such, 'in-itself', outside the sphere of representation) be 'given' to a subject? The evidence proper to the logician is sufficient for the constitution of a *Mannigfaltigkeitslehre*. But the profound meaning of concepts which he makes use of, remains obscure: this obscurity by itself explains the nearly universal confusion between logical and psychological laws which resulted when the *Logische Untersuchungen* made its appearance.

Thus, the explication of the meaning of logical objects led necessarily to problems regarding epistemology.[9] These problems require precisely a return to the lived. As we have said, when confronted by a logical law, we must not be satisfied with a 'vague comprehension'. This advice might seem astonishing, seeing that logical laws are rightly reputed as being the clearest of all. But the evidence can be sufficient for calculation; while, from a philosophical point of view, concepts remain confused. This confusion is prejudicial to logic itself, which becomes a simple technique without 'truth'. From this point, it is necessary to return 'to the things themselves', to effect in a fully conscious manner the acts of idealizing abstraction which render logical objects *present* to us. In this operation, 'logical evidence' is realized where the objects are 'given' to us. The phenomenological analysis of this evidence will allow us to specify the meaning of their being and to resolve the philosophical problems that they arouse.

It is appropriate to specify these indications, which may seem somewhat formal, according to the descriptions of the *Sixth Investigation*. The second part of it is devoted to the doctrine of 'categorial intuition', understood as the intuition of non-sensible objects that are intended in predication (*kategorein*): predicates, relations, states of affairs (*Sachverhalte*), etc.[10] A categorial object can be given 'in person', when we formulate the judgment according to the evidence; it also can be 'presented simply', when we speak about it absent-mindedly, without thinking about what we are saying. It is clear that in the latter case error is possible. Thus, the truth of science, as a system of propositions, is tied to the conditions of categorial evidence. And the description of this will be an essential task for the phenomenological theory of knowledge.

We do not have to get into the details of this text, which is confused at times and underwent many subsequent revisions, without ever succeeding in satisfying its author. It will suffice to retain its fundamental idea. In opposition to the sensible intuition, which as 'simple' intuition, needs nothing else in order to exist, the categorial intuition is necessarily 'grounded'. It presupposes, precisely, the presence, in the background, of a sensible (simple) intuition. The 'categorial acts', which are effected apart, can give but a vague and purely 'symbolic' comprehension. In other words, judgments formulated without the presence of a corresponding sensible intuition remain on the verbal plane: they are confused thoughts, removed from the *thing itself*. It is important to be precise here. The issue of a lapse into empiricism is not at all involved here: the 'thing' here is not the sensible object but, precisely, the *categorial* object defined by predication. But this categorial object can be present 'in person' only on the 'foundation' of the sensible object but nevertheless without this condition taking anything away from the originality of its being: for inasmuch as it is constituted 'on' the sensible, it will be defined as '*supra*'-*sensible.*[11]

We see from this example the meaning and scope of phenomenological analysis. Traditional theories of knowledge struggled with inextricable difficulties regarding the problem of the relations between the sensible and the intelligible. This ended with Kant's declaring that a concept without an intuition is empty and an intuition without a concept is blind. The connection was brought about through the mediation of the notion of 'possible experience'. With the categories defining the conditions of *objectivity*, any experience which would not conform to them would not be the *experience of an object*; consequently, it would not be an *experience* at all. Thus the *intuitive synthesis* which constitutes the objects of perception, proceeds from the unity of the understanding, which presides over the *conceptual synthesis*. In this manner, the possibility of a conformity between the objects *given* in perception and judgments *formulated* by science was founded – in other words, the possibility of knowledge. We shall have the occasion, later, of examining the Kantian doctrine more carefully. Here it will suffice to notice that the demonstration rested entirely upon arbitrary definitions, notably that of the notion of object. We find here a mixture of irreconcilable concepts. Is the object of *possible experience* a *sensible* or a *categorial* object? Does it refer to the object that I perceive this moment, or does it refer to an intelligible unity constituted in that network of necessary determinations "of which we have an idea when we speak of nature"? From then on, phenomenological analysis made this confusion impossible. The perceived object has a unique place in spatio-temporality. The categorial object as such is outside of space and time. That I am here at this moment, constitutes a true proposition, and, inasmuch as it is true, it is valid for all times and places. Though it is true that the subject is a particular being, yet the judgment intends a 'state of affairs' which as such constitutes a 'suprasensible' ideal object.

But phenomenology does not allow merely a full elucidation of the opposition between the understanding and the sensibility by the definition of their respective objects. *The analysis of correlative evidence resolves the problem of their relations*. The sensible and intelligible do not exist 'apart', since categorial evidence is always 'founded' on sensible evidence. But the concept of 'foundation' precludes any confusion between the spheres. It is not a matter of deriving the intelligible from the sensible, in the manner of the empiricists, since it belongs to the very essence of 'founded' acts to intend radically new objects. The intelligible is engendered by the sensible without being reduced to it.

In this way the value and limits of science are justified. The possibility of an accord between intuition and concept does not refer to the simplistic and arbitrary hypothesis of a construction of perception according to the categories of judgment, but to the very essence of categorial evidence as *founded* evidence. Predicative knowledge can be realized only in the presence of a sensible intuition, but it bears upon an object of a 'superior degree', characterized by its ideality. The propositions of science are true *of* experience, and are true only *inside of* experience. However, the result is obtained not by a confusion of spheres, but by an explication which implies their distinction. That the intelligible must be verified in the sensible does not diminish in any way the originality of their respective modes of existence.

Through the notion of foundation, phenomenological description allows for the correction of faults of eidetic analysis that could suggest a complete separation of types. The first volume of the *Logische Untersuchungen*, anxious to define the specific nature of logical objects, seemed to posit a world apart – a world of Ideas – and the author saw himself being reproached for 'Platonic realism'. But logicism has been avoided precisely by the 'subjective' investigation of the second volume, in which the analysis of categorial evidence demonstrates the necessity of a reference to the sensible – nevertheless without falling into the characteristic confusion of the Kantian notion of 'possible experience'. Thus, the return to lived experience was not a lapse into psychologism, but an effect of the influence of Brentano on the author. It was necessary to resolve the problem of knowledge through the explication of the very meaning of logical objects, inasmuch as they imply a relation to reality.[12] But the task was correctly seen possible only by means of the analysis of corresponding *evidence*.

The method implied that the description of lived experience was not a '*psychological*' task. It is clear that if the definition of categorial evidence as 'founded' evidence expressed only a peculiarity of the human psyche, the relation could not have any value on the plane of the object, where the intelligible and the sensible would remain definitely 'apart'. The proposed solution to the problem of knowledge had meaning only if the conditions of *the thing itself* corresponded to the conditions of evidence. In 1901 an unfortunate formula, still well-known, defined phenomenology as 'descriptive' psychology.[13] However, the latter was forcefully distinguished from empirical psychology, which looks for causal explanation,[14] and also from introspective psychology, based on the practice of internal perception. Here it is not the point to formulate a theory of 'psychic phenomena' alongside one of 'physical phenomena'. The object of the internal experience of the psychologist is *transcendent* just as it is for external experience: the intended transcends the *given*, and an error always remains possible. Phenomenological intuition deals with *pure immanence as such*: we accept it in the exact measure that it is *given* where doubt would be *meaningless*. Only this absolute evidence of the pure *phenomenon* can satisfy the exigencies of a theory of knowledge. The latter could not make use of ordinary evidence, which it puts into question, and rightly so: this does not mean that it must cast off all content in order to be content with a conceptual analysis in the manner of Kant but simply that whatever is not given in *phenomenological evidence* must be excluded.

In this way, phenomenological analysis, led on by considerations regarding a theory of knowledge was 'psychological' in name only. It is concerned with the absolute evidence of the consciousness of the self, distinct from internal perception. Thus, one can already speak of a *transcendental* analysis which should not be confused with the Kantian project. The study of the subject, in Kant, passed through the mediation of the object. The *a priori* forms of the sensibility and the understanding had been abstracted by an analysis of the *conditions for the possibility of an object*: their transcendent signification was immediately guaranteed, but it would be difficult to say why one would insist on attributing them to a 'subject'. The Husserlian investigation takes a diametrically opposite direction. Psychologism is not avoided by the refusal of a direct thematizing of the subject and the detour of an analysis of the conditions of the object: the transcendental dimension is reached precisely by the deepening of the consciousness of self, inasmuch as it reveals in *absolute evidence*, as *pure phenomenon*, the meaning of being and of the self.

Thus the passage to transcendental idealism will be but an explication. The *Logische Untersuchungen* contained it, implicitly. The first volume took the object as an in-itself; the second gave analyses of the 'real' content of the lived.[15] Thus, the author seemed to adhere to an ordinary realism, positing (on the one hand) the object as 'external' reality and (on the other hand) the consciousness which knows it. But examination of the actual analyses reveals the meaning of the work. The affirmation of the originality of the logical object was founded upon an intentional analysis of the aims of the logician. Logical laws are irreducibly distinct from psychological laws, because one does not intend the same things in each case. Conversely, the phenomenological analysis of the last two *Investigations* seemed to enclose itself in the 'real' content of consciousness. But the first characteristic that we find there is defined as a surpassing – consciousness is consciousness *of* something; in itself it requires something other than itself: the object that it intends. Thus, the two volumes of the *Logische Untersuchungen*, if we take them in their *truth*, constitute the two moments of an authentic phenomenological analysis in the classical noetic-noematic form. The first moment describes the *noema*, the object intended by the logician inasmuch as it is distinguished radically in its ideality from all psychological reality. The second moment describes the *noesis* which intends it. But this correlation was only 'in-itself' or 'for us'. As a 'for-itself', it took on a form which contradicted its real signification. To be sure, the intended object has no other meaning than through the act of consciousness. But in the naiveté of a direct description, it is posited as an independent entity, since the logician intends it precisely as such. Conversely, consciousness was, from that time on, a constituting consciousness, inasmuch as it gives its meaning to the object. But the first look of the phenomenologist takes it only in its 'real' content, as a thing enclosed within itself.

In this manner, the 'realism' of the *Logische Untersuchungen*, ignorant of its own meaning, was but the contradictory form of an intentional analysis. It implied the necessity of a dialectical reversal. But in the Husserlian language of the last period, one would say that the profound intention of the work *strove for* an idealism, but that it was *actually expressed* as a realism. But if we use Hegelian vocabulary, then we will have to reverse the terms and say that the author *intended* and believed that he had posited a realism, but that he *attained, in fact*, the relation of constitution. In that 'experience', philosophical consciousness is going to take a new 'form' – that which will be expressed in transcendental idealism.

6. THE DISCOVERY OF THE REDUCTION

Husserl taught the theory of the reduction for the first time in the first five lectures of a course entitled '*Fundamental Points of Phenomenology and of the Critique of Reason*' at Göttingen during the summer semester of 1907.[1] Through that text we are going to try to clarify the original signification of the doctrine.

The author begins with a certain number of commonplace considerations about the theory of knowledge. Natural thought develops on the plane of the object; the existence of the world and the possibility of knowing it does not raise any difficulty for it. But what is taken for granted by the ordinary man becomes a mystery for the philosopher. Knowledge, indeed, is understood, as a fact alongside of many other facts – a psychological event. How then can it arrive at a truth? In the naturalistic interpretation, one that is inevitable on the plane of naive consciousness, every value would disappear. Psychologism, which is adopted spontaneously by thought when it wants to understand itself, results in skepticism. Reflection on the problem of knowledge requires surpassing of the natural attitude.

Such is the content of the first lecture. We recognize well-known themes in it. To transform the *subject* into *object* is to make it incomprehensible, and this is precisely what one does when philosophizing with methods familiar to natural thought. Everyone knows the classical solution: in order to avoid the peril of psychologism, it is necessary to proceed to the conceptual plane by a reflection on the 'conditions for the possibility of'. But we will ask why, in this case, should one insist on speaking of a 'subjectivity' – albeit a 'transcendental' one. Husserl's merit was in maintaining the necessity of remaining on the concrete level. The exclusion of objects of nature does not condemn us to be enclosed in a system of forms. All that is needed to give philosophy its own content is the discovery of a mode of superior knowledge.

Such is the goal of the second lecture. The problematic of knowledge had transformed the obviousness in the natural attitude into mysteries. The philosopher must not take into account any of the propositions which he has admitted up until now. He must find a new kind of knowledge which is absolutely certain. For that, it will be enough for him to work his way through the Cartesian method. It is possible that the object may not exist, but consciousness is given to itself in the clarity of the most absolute.

> *Every lived process whatever*, while being enacted, *can be made the object of* [intuition] *a pure seeing and understanding, and is something absolutely given in this 'seeing'*. It is given as something that is, that is here and now, and whose being cannot be meaningfully doubted.[2]

It is true that the mode of this being must still be specified. But one point remains certain: we have reached the terrain of the absolute: the certainty that consciousness takes of itself in its accomplishment. "This perception is, and remains, while it lasts, something absolute, something here and now, something that is in itself what it is, something by which I can measure, as by an ultimate standard, what 'being' and 'being given' can and must mean."[3]

Thus, *the problem of the critique of knowledge found its solution in a return to the Cartesian cogito, as a temporal and actually lived cogito.*[4] Realism, which belongs to the natural attitude must be discarded by a 'reduction' required by the theory of knowledge (*erkenntnis-theoretische Reduktion*)[5] which suspends all transcendental positions. But the absolute domain of concrete immanence is not affected by this, for in it will be found the final solution to the problem of cognition.

This synthesis of Kantianism and Cartesianism might appear contradictory.

Does it not involve falling back into psychologism after having reached the transcendental plane? A new lapse into the natural attitude, responds the author at the beginning of the third lecture, is undoubtedly always possible, and even inevitable. The pure given of lived immanence is understood spontaneously as an event in the world. So, to protect phenomenology from every psychologistic interpretation, it is necessary to perform anew, on the immanent given, the reduction required by the theory of cognition. Seeing that the reduction now makes explicit the absolute meaning of 'pure phenomenon'[6] and distinguishes it in a radical manner from the 'psychic phenomenon', this reduction deserves the name of 'phenomenological reduction'.[7]

Thus, the *phenomenological reduction* specifies only the meaning of the domain that was thematized as a prelude to the Cartesian *cogito*. This last step is itself preceded by a first *reduction proper to the theory of cognition*. The three steps succeed each other without revealing any intimate connection between them. The exclusion of objects of the world – the '*erkenntnis-theoretische Reduktion*' – refers to the critique of knowledge. In Kantian language, the 'conditions for experience' could not depend on 'experience' itself. But the procedure remains strictly negative and cannot, by itself, provide us with any content. The lived given is thematized only by the Cartesian movement which seems to be juxtaposed to the first in an arbitrary manner. Finally, as if the author perceived, after all, the incoherence of his procedure, he returns once more to the reduction in the domain thus revealed by qualifying it this time as 'phenomenological'. But these obscurities should not make us misjudge the value of the text. The authentic signification of a philosophical idea is revealed in the very moment of its discovery, where the concern for conceptual coherence has not yet masked the lived inspiration.

First of all we will carry out our examination of classical references. Husserl did not simply reproduce the Kantian and Cartesian movements by juxtaposing them. If the reduction finds its unquestionable historical origin in the works of Descartes and Kant, it nevertheless has its own intrinsic meaning which, in turn, will shed light on these works.

In beginning with *The Critique of Pure Reason*, we will point out that the original concern was not, as it will be the case of phenomenology, with knowledge in general, but more precisely with *a priori* knowledge. "How are synthetic *a priori* judgments possible?" One recalls the answer: they are possible as conditions of synthetic *a posteriori* judgments, and only as such. In this manner and at the same time, the *value* and *limitations* of *a priori* knowledge were specified; the possibility of a mathematics and a pure physics and the impossibility of a metaphysics.

This demonstration was made by a return to the subject as bearer of 'possible experience'. Since every object is an object of experience *for a subject*, the *a priori* forms of the subject will condition the structure of the object. We might believe that these forms are extracted by a direct study of consciousness. But we know that this is not at all the case: the analysis only bears, in fact, on the *object* whose structures it defines: spatio-temporality, quantity, causality. These structures are transposed onto the *subject*, who guarantees their *aprioristic* value: inasmuch as they are posited as categories of the understanding and forms of the sensibility, they will apply necessarily to the object, since an object that would not conform to them would be nothing *for consciousness*.

We see immediately the meaning of the return to the self in Kant. *It was not a question of revealing the lived as such, but of finding in the concept of the self a point of reference which would allow for the foundation of the aprioristic value of the pure theory of the object* (mathematics and pure physics) *and at the same time would limit its range to the domain of possible experience. The transcendental self, entirely empty in itself, will have no other content than that very same content extracted by an analysis of the structure of the object.*

Thus, the thematizing of concrete consciousness was precluded by the original motivation. *What was important was the justification of a system of synthetic a priori judgments that were constituted already.* The return to the self was not done for its own sake but solely for that justification. Such a function could be fulfilled if only the transcendental self be defined originally as an empty relation to the object – a relation which finally would enable us to attribute to a consciousness, inasmuch as it is a consciousness *of the object*, the corresponding forms as the conditions of the object.

The prohibition against thematizing concrete consciousness had its *de jure* foundation in the well-known doctrine of the ideality of time. The temporal form, which interposes itself between the self *as it is* and the self *as it appears* to itself, definitively removes all philosophical value from the project of attaining an actual knowledge of the self.

From this point of view, the Kantian system has realized *in toto* its historical goal: to establish for *a priori* knowledge, at one and the same time, both its value and its limitations. *But there remained the problem of knowledge itself*, in its concrete content. If Kantianism, after having established the general conditions for the *possibility* of experience, did not give any valid answer to the well-known question of how the actual data of *real* experience could accord with *a priori* 'form' in order to allow for the constitution of *particular* laws of nature, the reason for this lies in its origin: namely, that the system had been constructed for mathematics and pure physics, and not for physics, properly speaking. More precisely, it was a question of spurning the transcendental pretentions of rational thought by giving pure concepts a foundation which would limit their usage to the domain of experience. The anti-Leibnizian controversy is the historical motif of the limitation of critical reflection to the problem of synthetic *a priori* judgments. But once the goal had been achieved, a demand to go beyond it had to be faced. After Kant, no one could believe any longer that the absolute could be known with the concepts of the object. The polemical interest was dying out with the disappearance of the opposition. In return the system, because of its internal logic, revealed the problem of *a posteriori* judgments.

As universal laws of nature have their ground in our understanding, which prescribes them to nature (though only according to the universal concept of it as nature), particular empirical laws must be regarded, in respect of that which is left undetermined in them by these universal laws, according to a unity such as they would have if an understanding (though it be not ours) had supplied them for the benefit of our cognitive faculties, so as to render possible a system of experience according to particular natural laws.[8]

This involved reverting to finality, that '*asylum ignorantiae*'. Kantianism did not provide the necessary means to resolve the problem that it itself had made inevitable. Empirical knowledge, which is the only actual knowledge, could not be thematized clearly, except by means of a concrete method. It was no longer necessary to take the self in its empty relation to a possible object, but as the subject of a real experience. The procedure required that the dogma of the ideality of time, whose historical reason we have noted, be renounced; for, it is opposed now to the analysis of the actual subject which is, to be sure, a temporal subject. Already, brilliant investigations by contemporary critical philosophers, which were done in the classical fashion by a reflexive analysis not of the act itself but of its products in science, had imposed the idea of a *becoming within the categories*. Such a position rendered untenable the bastion of the Kantian fortress, which was the ideality of time. The Husserlian problematic, by directing the theorist of knowledge toward the analysis of lived consciousness, does not constitute a lapsing into psychologism, but the explication of the *result* of a dialectic immanent to criticism itself. The pure concepts which defined a nature in general, found an adequate 'vehicle' in a formal self, as the simple 'condition for the possibility' of empirical consciousness. But these pure concepts, owing to their meaning, referred back to actual knowledge, that of real objects, a knowledge which can be referred only to concrete subjectivity. This concrete subjectivity rightly must be defined to the exclusion of all *worldly* concepts in order to satisfy the demands of philosophical thought: but this exclusion does not have any bearing on its *concrete* character. The 'reduction proper to the theory of cognition' does not reproduce the Kantian movement which was motivated by the demand for an empty self, the simple bearer of 'conditions for the possibility of'. Husserl's objective was to eliminate the transcendental positions in order to thematize the actual act whereby they find their signification: the problem of knowledge will find its solution in a *transcendental experience* – that of the Cartesian *cogito*.

These considerations lead us to compare the Husserlian method with its precedent in Descartes. Our task will be facilitated by a remarkably lucid passage of an unpublished course on the *Idea of Phenomenology and of Its Method.*[9] *The weakness of Descartes*, observes the author, *consisted in his not having started with the difficulties of the theory of knowledge*. Also, for him philosophical inspiration was combined with the desire, typical of the natural attitude, to arrive at a universal science, in the sense of the exact sciences. The project of a *Mathesis Universalis* kept him within the horizon of mechanism, while the struggle against skepticism bore witness to his aspiration to absolute knowledge, in the philosophical sense. Also, the *cogito* (which he, nevertheless, had lived authentically) relapses with the *sum*[10] into the natural world. [*Cogito ergo sum*, an actual lived thought.] As for us, we take as our starting point the problem of knowledge. The being *in itself* of the object is revealed as an enigma: we question it, but without contesting it as does Descartes, who follows the skeptic on his own grounds. "What he seriously places in doubt, we ourselves call into question." We have only to extract "what has not been treated by being called into question." Descartes says: "It is possible that the perceived does not exist, but the perception of it is indubitable." But we say instead: "The perceived is in question, but the perception is not."[11]

It seems to us that these remarks shed light on the problem. Descartes remains on the plane of being. The opposition between the doubtful and the certain is internal to the uncritical naiveté typical of the natural attitude. The consequence of this is seen immediately: transcendental existence to which he had raised himself, thanks to the *cogito*, is interpreted immediately as natural existence: *sum res cogitans*. The self is a substance which will insert itself in the world when the world will have been rediscovered as a consequence of divine veracity. But how can a being of the world know the world which is outside of it; or, more precisely, how can a representation, as an event in nature, lay claim to a truth value? That was the problem which, not having been posed from the start, rendered the system untenable. The Husserlian *cogito* that in its being lived was identified with the Cartesian *cogito*, is interpreted in radically different concepts. It is no longer a matter of the worldly opposition between the doubtful and the certain, where one certifies for the latter the existence which one contests for the former: it is a question of distinguishing two radically different meanings of being.

"What Descartes seriously places in doubt, we call into question." The world is there. In a sense, its existence asserts itself. At any rate, we won't dispute it, as we would dispute a proposition insufficiently demonstrated. But reflection reveals an enigma in this: how can the *in itself* be *for me*? The existence of the world has not become doubtful but incomprehensible, and no object of the world could deliver to us the key to the mystery, since they all participate in its unintelligibility. But the sphere of pure immanence that is discovered by the *cogito*, precisely escapes this difficulty. The lived as such is not *in itself* since all of its being consists in being *for me*: an absolute that we cannot 'call into question'. One could not very well, then, interpret it from the standpoint of the natural attitude, which consists precisely in treating objects as things *in-themselves*. It is not a matter of a part of the world which is certain, in opposition to the rest that is declared doubtful; it is a matter of a radically new mode of existence which is not treated by the general 'calling into question' which motivated the mystery of worldly knowledge.

Descartes, starting with the problem of skepticism, at least remained within his concepts, on the plane of the naive attitude: what he salvaged from doubt he attributed precisely to the very mode of existence contested by the skeptic; namely natural existence. Husserl, starting with the problem of cognition, does not seek to arrive at a (worldly) certainty, but seeks to find the very meaning of all certitude. Thus, the *cogito* does not reveal consciousness as an object of the world salvaged from universal doubt: it is an absolutely original domain, which is posited outside of all worldly existence.

Thus, the third step which we have extracted from our analysis of the text of 1907, the *phenomenological reduction*, does not constitute a sort of correction, added as an afterthought, in order to escape the reproach of psychologism. The Husserlian *cogito*, motivated by the demands of its theory of knowledge, revealed from the beginning the domain of a non-psychological immanence. The phenomenological reduction, by insuring that lived experience not be given any realistic interpretation was only making explicit the meaning that it already possessed at the very moment of its thematization: it marked the radical originality of what we had been tempted to consider a simple juxtaposition of Kantianism and Cartesianism.

We have specified the meaning of the phenomenological movement with respect to its original accomplishment. It still lacks an actual realization. The phenomenological domain is clearly defined as non-psychological: it still is necessary to demonstrate this as such in a positive exploration. What truths can we enunciate about the experience at which we have just arrived?

The problem of the method of description entered into in the third lecture confronts us immediately with a fundamental difficulty. The Cartesian evidence hands us a flow of lived experiences that are always new: how can we express propositions regarding this *Heraclitean flux*? It is obviously necessary to go beyond the individual phenomenological data. Every judgment introduces universal notions. How then can we maintain that the principle of evidence is to cling to the 'given' so far as it is given?

It is sufficient, replies the author, to dispel a misunderstanding. Phenomenological experience does not bear solely on the individual features of the 'real' course of consciousness, but also on universal ones: the essence is part of the pure given. The fourth lecture explains this by means of an example. I perceive a red. I perform the phenomenological reduction.

I prescind from any further significance of red, from any way in which it may be viewed as something transcendent, e.g., as the redness of a piece of blotting paper on a table. I fully grasp in a pure intuition the *meaning* of the concept of 'red' in general, redness *in specie*, the identical universal that intuition detaches from this or that. No longer is it the particular as such which is intended, nor this or that red thing, but redness in general.

We then have a pure immanent consciousness of the universal, an 'Immanent given'.[12] The essence of cognition may be grasped in a similar manner.

This explanation seems misleading to us. The essence, as an essence *of* the world, will be part of this world: it is transcendent, not immanent, and this will be the very doctrine maintained in the *Ideen.*[13] On the other hand, if we look at the example that has been presented above, we see that, after all, it is just a conceptual analysis. We abstract 'the meaning of the concept of redness in general'. No one will be surprised that these results enjoy the benefit of absolute obviousness: all we have done is to explicate the content of a concept. But we do not see that we have made any progress in our knowledge of the thing.

We have shown in the preceding chapter the ambiguity of the notion of essence, inasmuch as it is interpreted sometimes as *pure possibility* and at other times as *possibility of the real.* The danger will become patently clear if we consider the problem of knowledge. Is it a matter of explicating the content of the concept that we have of knowledge, or rather of defining knowledge in its authentic essence? In the former case, the analysis is relatively easy, but it runs the risk of sanctioning social prejudices. Each era has a different notion regarding knowledge; and the philosopher, owing to the illusory necessity of conceptual analysis, constructs an eternal truth from it. Undoubtedly, this was not the intention of the author. His real concern is to abstract the *true signification* of knowledge, that without which it would not be knowledge, because it constitutes its very *essence*. But an analysis of this kind can hardly remain on the plane of the simple concept. There has to be a return to the actual lived in order to explicate the meaning which constitutes its very being. In the *experience* of authentic knowledge is revealed the very thing which constitutes its truth value.

We see now in what sense it is appropriate to understand this affirmation that the universal enters into the domain of immanence. It is not a matter of discovering, in addition to the evidence for the particular data of the stream of consciousness, an evidence no less absolute which would bear on universals. Such an extension of the phenomenological domain would be meaningless. It absolutely cannot be a question of departing from the Cartesian evidence for particular *cogitations: it is in the very actuality of their being lived that it is possible to discover their eternal meaning*, provided that the procedure that revealed them had at least a *raison d'être*. The phenomenological reduction would be entirely useless, if, after having discovered the absolute domain of concrete immanence, whose certainty resides in the very act by which it is actually experienced at each instant, it had to be abandoned under the pretext that there is no science of the particular. To say that universals also benefit from the phenomenological evidence, is to say that a universal meaning is found that is immanent to each moment of the *Heraclitean flux* of subjective appearances. And if the Cartesian *cogito* has been recognized as necessary for founding the theory of cognition, it is so because this meaning can be discovered only as a result of the realization of the particular.

The text on which we have just been commenting, offers us the meaning of the doctrine, in the authenticity of the original act. The issue at hand was to resolve the problem of transcendence – of understanding how consciousness can surpass itself so as to take on a value for eternity. The classical solution consisted in realizing this value in the ideal forms or in a transcending being in which the self participates with its understanding and will. Thereby, the actuality of subjective existence, in its lived intimacy, is found to be irremediably sacrificed. The self can go beyond itself only by abandoning itself to itself, and the exigencies of concrete consciousness are relegated to 'psychology'. Husserl's originality consisted in looking for the solution to the problems of rationalism in the actual act of individual subjectivity. Spiritual life is not an abandonment of the self in the cause of superior norms; rather, it pertains, to the essence of consciousness to transcend itself, and the eternity of the idea possesses no other meaning than as a result of its lived act.

7. THE EXPOSITION OF THE *IDEEN*

The exposition of the Reduction, set forth in the second section of the *Ideen*, presents some rather important differences from the lectures of 1907. The suspension of transcendent theses is no longer motivated by the exigencies of the theory of knowledge; it is proposed simply as a possible decision for freedom. On the other hand, the certainty of immanent data is no longer admitted as self-evident: it calls for a demonstration which constitutes the most celebrated part of the work.

We start with a naive description of the *natural attitude*. The world is there before me as an ensemble of spatio-temporal objects of which I myself am a part. It is always in this totality, in virtue of a "dimly apprehended depth or fringe of indeterminate reality"[1] that I perceive particular objects. Every object is an object *in the world*, and this expression points to practical possibilities of progressive discoveries: I can enlarge my actual perception; I can evoke in memory past events – all operations which are always limited in themselves but which constantly presuppose a *horizon* of infinite possibilities: "The zone of indeterminacy in infinite. The misty horizon that can never be completely outlined remains necessarily there."[2] To be sure, the world is not simply a world of things; for, these things are bearers of predicates of values and ends. Some of these are similar to me myself: these are animals and other men. But all of these are integrated by me within the horizon *of one and the same world* – the world which is there and with which I am constantly concerned.

Instead of "freely disposing myself" for this spontaneous life, I can, by an *act of freedom*, effect a radical change of attitude. I can *suspend* the general status of existence, by no longer believing in the world, even while living in it. Let me be precise here. It is not a matter of doubting this or that, since every particular act of doubt is effected within the general horizon of the world posited as existing already here and now. Nor is it a universal doubt like that of Descartes, which involves the supposition of the non-being of the world. It would be even less a matter of a sophistic negation of the existence of reality. In a word, in no wise do we take a position, but we content ourselves with *abstaining from taking any position*, all the while continuing to live our natural lives. We continue to believe in reality, while refraining from participating in this belief. The world is always there.

We make no change in our conviction, which remains in itself what it is . . . . And yet the thesis undergoes a modification – whilst remaining in itself what it is, *we* set it, as it were, '*out of action*'; '*we disconnect it*', '*bracket it*'. It still remains there like the bracketed in the bracket . . . . The thesis is lived, *but we make 'no use' of it . . . . This transvaluing is a concern of our full freedom . . . .*[3]

Such is the phenomenological *epoché*.

But a question presents itself. What can remain after this universal 'suspension'? We have 'disconnected' not only the 'external' world but ourselves and every type of *cogitatio* as psychological reality. But perhaps there is simply an illusion which makes us interpret the lived acts of consciousness as real events in the world. Thus we shall retrace our steps and "without having yet carried out the phenomenological suspensions of the element of judgment, we subject consciousness to an *eidetic* analysis."[4] Thus we shall see "*that consciousness in itself has a being of its own (Eigensein) which in its absolute uniqueness of nature remains unaffected by the phenomenological disconnection*. It therefore remains as a '*phenomenological residuum*', as a region of being which is in principle unique, and can become in fact that field of a new science, the science of phenomenology."[5] It is only then that the *epoché* will deserve the name of *phenomenological.*

So, let us return to the natural attitude. We immediately find ourselves facing two spheres of existence: the 'external world' of things and lived consciousness, which is always given to itself and explicating itself in *reflection.*[6] The world is present to consciousness, and each lived act is directed to the world. But no one can confuse the object itself with the consciousness which intends it. The object is intended, and rightly so, as a reality which does not belong to lived experience: perception is always *transcendent* to it, inasmuch as it necessarily must go beyond itself in order to grasp it. On the contrary, consciousness is presented constantly to itself in a lived relation of self to self: it can explicate itself only through an *immanent* perception in which, reflecting upon itself, it forms an immediate unity with itself. Thus we can distinguish in a radical manner two types of relations: the 'real' (*reell*) relation of lived moments within a single stream of consciousness, and the *intentional* relation by which consciousness intends a *reality* (*real*) as other than itself.

By now it is quite clear that no *reality* as such can 'really' (*reell*) belong to consciousness. For the sake of clarity, we shall limit ourselves to the case of sense perception, which we can consider as the primary experience of the world: the *sensible material* world is the fundamental stratum upon which all given objectivity in the 'natural attitude' is constituted.

The *thing*, the object of sense perception, is given as a unity identical with itself throughout an incessant stream of 'silhouettes' (*Abschattungen*) in which it 'appears in profile'. These silhouettes are the lived sensations in me and are animated by a 'sense of apprehension' which relates them to the object. The thing which 'appears in profile' in them is precisely and nothing other than the unity intended throughout the multiplicity of these modes of appearance. The sensible qualitities themselves, as real moments of the thing, are grasped only through a stream of ever-changing *silhouettes*. Because of them, there is no confusion possible between the lived experience that I possess at every moment *in myself*, and the object itself which is presented *to me*. "The silhouette is a lived experience. But the lived is possible only as lived, and not as spatial. On the other hand, the object which appears in profile in these silhouettes is possible only as spatial (it is spatial precisely in its essence); it is not possible as lived." [7]

Thus we arrive at a distinction that is absolutely fundamental: "that of the *being as lived* and of the *being as a thing*." The thing, offering itself only through *profiles*, will always be *transcendent* to the consciousness which intends it. But, the lived does not give itself through profiles: it is grasped in itself as a 'real' (*reell*) moment *immanent* in consciousness. It is clear, however, that a perception through profiles involves a certain persistent inadequacy. The thing is seen at every instant in only *one of its aspects*, which refers back to a relatively indeterminate horizon of possible perceptions. Undoubtedly, the very meaning of the object implies a certain *style* of development. But the actual course of perception will bring always new and unforeseeable data – corrections revealing that what was thought to be real is shown as illusory.

On the contrary, immanent perception grasps the lived as an absolute: the lived has no 'aspects' which could be presented sometimes in one way and at other times in another; it is precisely as it presents itself in the lived consciousness of self. Consequently, the existence of the thing is always in doubt, since all transcendent perception can be exposed in the course of experience as a simple error. The lived, on the contrary, could not be denied inasmuch as it is lived.

The position of the world, which is a 'contingent' position, is opposed to that of my egological life, which is a 'necessary' position that is absolutely indubitable. Every given thing 'it-self' ('in person') *also can not be; any lived given 'it-self'* ('in person') *can not not be.*[8]

We can now come to a conclusion. We were looking for the 'phenomenological residue' which had to remain after the general suspension of the status of the world. We now see that it is concerned with lived consciousness, which, inasmuch as it is lived, possesses an absolute being independent of the being of things. In fact, the world is revealed as the transcendent unity intended in the movement of actual experience. Its existence is only the correlate of a specific structure of the lived in this movement. Let us now imagine that the stream of consciousness takes on a completely disordered form: it is conceivable that experience be dissolved into appearances because of internal conflicts:

that experience shows itself all-at-once obstinately set against the suggestion that the things it puts together should persist harmoniously to the end, and that its connectedness, such as it is, lacks the fixed order-schemes of perspectives, apprehensions, and appearances – that a world, in short, exists no longer ... it is then evident that the *Being of consciousness*, of every stream of experience generally, *though it would indeed be inevitably modified by a nullifying of the thing-world, would not be affected thereby in its own proper existence .... Thus no real thing*, none that consciously presents and manifests itself through appearances, *is necessary for the Being of consciousness* (in the widest sense of the stream of experience).

Immanent Being is therefore, without doubt, absolute in this sense, that in principle it has no need of any 'thing' in order to exist (nulla 're' indiget ad existendum).

On the other hand, the world of the transcendent thing (res) is related unreservedly to consciousness, not indeed to logical conceptions, but to what is actual.[9]

Thus, the *epoché* is not simply a negative procedure: it reveals the concrete life of pure consciousness in its absolute being. In virtue of intentional unities, acts of transcendence are constituted within the stream of experience. Thus, inasmuch as consciousness contains the solution to the problem of transcendence, it deserves the name of *transcendental consciousness*.

The suspension of the status of the world was, in 1907, called the 'reduction proper to the theory of knowledge'. The *epoché* is posited in the *Ideen* as a simple act of freedom. Nevertheless, the critical problem remained an underlying one for the whole development. It appears in Section 33 where 'pure' consciousness is defined as 'transcendental' consciousness, and especially at the end of the exposition of the reduction, at section 62, where the opposition between the dogmatic and phenomenological attitudes refers expressly to the traditional opposition between dogmatism and criticism. We will ask ourselves with all the more astonishment why the author did not return to the remarkable motivation of the original text. It is precisely the explanation of this particularity which will provide us with the meaning of the *epoché*.

The text of 1907 reproduced a classical procedure. The status of the world was excluded by reasoning about the *consequences* of naturalism: every interpretation of the act of knowledge as an event in the world renders its truth-value unintelligible. The argumentation took place on the *discursive* level, the *concept* of knowledge being incompatible with its integration with empirical reality. Hence, the 'reduction proper to the theory of knowledge' imposed itself with an *analytic necessity*: but it could be realized only in the *conceptual* mode. The transcendental positions were eliminated by a procedure, which, since discursive, remained strictly negative. It could not by itself disclose any new content. Consequently, the Cartesian return to concrete experience appeared to be some kind of contradiction: it would have been, so it seemed, more logical to study the subject of knowledge in the *concept* by the determination of the conditions for the possibility of the object.

But it belongs to the essence of phenomenological reflection to proceed by means of intuition. For Husserl, the suspension of the status of the world could not retain the discursive aspect which characterizes it in critical philosophy: it could be only a matter of a lived procedure, and the absolute domain of concrete immanence found itself thereby immediately thematized. The *epoché* must reveal itself precisely as the 'necessary operation' which makes it accessible to us.[10] It is true that the exposition of the *Ideen* then turns back to proceed to an eidetic analysis of consciousness on the plane of natural naiveté and ends with the conclusion that pure experience has a being of its own which makes it appear as the *phenomenological residue*. But this in no way means that the *epoché* had been purely negative; it was simply a matter of demonstrating that the domain which it had disengaged indeed corresponds to its concept.

Natural life is not defined solely in terms of the status of objects in the world. Every consciousness is consciousness *of some thing*, but the consciousness of the object as such is always from the outset, already *consciousness of self*: that very same fact defines it, and rightly so, as *consciousness*. Undoubtedly, what is lived is perceived, properly speaking, only in a *reflection*. But this reflection

> . . . has this remarkable peculiarity that what is perceived in it is characterized as, in principle, something which not only is and endures within the gaze of perception, but *already was before* this gaze was directed to it.'All that is lived is conscious experience' tells us specifically with respect to intentional experiences that they are not only the consciousness of something, and as such present not merely when they are objects of a reflective consciousness, but that when unreflected on they are already there as a 'background' and therefore in principle, and at first in an analogical sense, *available for perception*, like unnoticed things in our external field of vision.[11]

Thus the apparent novelty of the reflexive content is due solely to the exclusive direction of attention toward the world, viewed from the natural standpoint: experience was already there in the *background*. The suspension of the status of the world has, as its immediate result, a topsy-turvy scene: the gaze of consciousness turns upon itself and explicates the content which, previously, was only *experienced* and not *perceived* as such. The *epoché* is not a discursive procedure but is itself a conversion that is actually experienced. It is not concerned with pronouncing an exclusion that is dictated by the analysis of the *concept* of knowledge, but with *abandoning the world in order to return to the self*. The bracketing of the world lifts away the *veil* which was hiding the self from itself. In this *unveiling*, the self continues to remain itself, as it was already present to itself in the natural attitude, and as it recognizes itself in an explicit consciousness of self.[12] The concrete content of subjective life does not disappear in the passage to the philosophical dimension, but reveals itself there in its authenticity. The position of the world has been 'put out of action', but not annihilated: it remains alive, although in a 'modified' form, which enables consciousness to be fully conscious of itself. The *epoché* is not a logical operation that is required by the conditions of a theoretical problem: it is the procedure which gives access to a new mode of *existence* – transcendental existence as absolute existence.[13] Such a signification can be realized only in an act of *freedom*.

It remains on the doctrinal plane, for one to demonstrate the legitimacy of concepts, and it is here that difficulties begin to arise. Does thematized consciousness in the *epoché*, answer to the conditions that are posed? The lectures of 1907 accepted the Cartesian *cogito* as self-evident. In 1913 the author confronted the need for a justification. We return to the natural attitude and perform a 'psychological reflection'.[14] And it is within this framework that the eidetic analysis of the lived data allows us to attribute to it a being of its own which resists the *epoché*. This 'zigzag' procedure may arouse some suspicion. Why return to the world after having left it? A more serious problem: if the opposition between consciousness and its objects takes place originally at the heart of the natural attitude itself, has anything more been accomplished than breaking in two of the total domain of worldly existences? What use is the argument if it only makes an absolute out of the psychological? Moreover, we know that there are *no differences in content between psychology and phenomenology*. "To each eidetic or empirical determination on the one side there must correspond a parallel feature on the other."[15] The propositions of transcendental phenomenology do not, it is true, bear any relation to nature, whereas the psychological inserts itself in the world through the intermediary of the body. But since the content remains strictly the same, has any more been accomplished in assuming the transcendental attitude than working through an arbitrary abstraction? The elimination of every relation to the body does not seem apt to provide anything other than a *descriptive psychology*.

Hence, it is not surprising that we learn that 'pure' consciousness – i.e., the ensemble of lived experiences as such – deserves the name of 'transcendental', because in it is found the solution to the problem of knowledge. This passage is made by the theory of the material 'thing' (*Ding*). The analysis of sense perception had revealed its object as the ideal unity intended in the indefinite stream of subjective syntheses. This result takes on an ontological import: the being of the 'thing' exhausts itself in the unfolding of 'silhouettes' (*Abschattungen*)[16] by which it appears in profile for us. There is no *in itself* behind the appearances: the thing *is* nothing but that selfsame thing which thus appears. The *real* object is reduced to the *intentional* object. Its laws are reduced to the synthetic forms of its modes of appearance, and objectivity is defined in terms of the organization of sensible data which are subsumed under an *a priori*. This proposition is valid for the world in general, with the constitution of the 'thing' having the double value of foundation and example. Transcendence has no other meaning than as transcendence *for me*, and its whole *being* consists in *being given* in the infinite system of appearances in which it is *constituted*. The world is only the *intentional* correlate of consciousness. Consequently, the annihilation of the world leaves intact the being of lived experience. More precisely, the world annihilates itself in the light of analysis: its whole being is reabsorbed in the intentionalities of subjective syntheses.

Thus, access to the phenomenological domain is made possible by the discovery of the ideality of the object. Evidently, what we are dealing with is a *transcendental* ideality inasmuch as this is identified with *empirical reality*. The foregoing procedure does not consist in dissolving transcendence into immanence (which would be lapsing into psychologism), but rather in explicating the very meaning of transcendence. There is nothing more in the unity of the object than the unity of the meaning; to speak of the absolute reality of the object is the same as speaking of a 'round square'. Transcendental idealism no more diminishes the value of objectivity than the denial that the square is round touches on its being.[17]

Phenomenology owed its first successes to the realism implied in the analysis of essences. It was a matter not merely of affirming an *in itself* of the universal. The *Wesensschau* claimed to reach *the very essence of things, the being of the existent* in all its possible forms. This thesis was asserted rather than demonstrated, but Husserl transmitted his conviction by the living example of his thought. The immense horizons that it discovered contrasted with the cold intellectualism of the critics in the universities. To be sure, there was no lack of 'subjective' analyses, but they did not seem incompatible with the status of an *in itself*. The ontological theory of the *thing itself* was being brought to completion by the description of the lived experiences in which it was *given*. The doctrine of the reduction, presented for the first time in his lectures in 1907, still had not explicated its fundamental presupposition – the ideality of the object. Consequently, those who attended his lectures hardly noticed it. The publication of the *Ideen* in 1913 provoked general amazement. Here transcendental idealism was connected to the very text of the *Critique of Pure Reason.*[18] "The term 'object' is always for us, a name for the *a priori* forms of connection within consciousness."[19] Phenomenology is defined as an "eidetics of a consciousness that has been purified by the transcendental reduction": the thematized field was concrete, but it is studied solely in its essence. So, seeing that the objectivity of the object was defined precisely in terms of the *a priori* of subjectivity, this was merely a recapitulation of the worn-out themes of critical philosophy.

His disciples violently protested in the name of the oral tradition. The doctrine's new orientation appeared to them as an abandonment of the authentic phenomenological inspiration.[20] Husserl saw his prestige growing every day, in university circles and abroad: he would never again know this warm atmosphere of enthusiasm which stimulated interest in the school of Göttingen.[21]

8. THE CRITIQUE OF THE KANTIANS

The critical philosophers, at first, favorably welcomed the new doctrine. Natorp immediately declared that he subscribed to it in its basic thrust.[1] The classical phenomenological themes of the intuition of essences, developed in the first section of the *Ideen*, ended unquestionably in an intellectualist realism reminiscent of Aristotle and scholasticism. But the constitution of the object, developed beginning with the second section, is in agreement with the standpoint of critical philosophy. Being is not a ready-made given but the ideal unity of subjective syntheses: it designates an infinite task to be accomplished in the movement of knowledge, an 'idea' (in the Kantian sense). Undoubtedly, Husserl does claim to define concrete experience as 'transcendental': this was to return to psychologism. One cannot see how a simple 'bracketing' could give us the absoluteness of consciousness. Subjectivity can be rediscovered only by a difficult task of 'reconstruction'.[2] However, the actual analyses of the *Ideen* once more take on an 'objective orientation' which reestablishes the critical tradition. *Intentionality* corresponds to the critical philosophy's concept of *objectivation*, and the analysis of consciousness constantly refers to the object which it intends. Husserl claims to thematize the 'real' content of consciousness. But the distinction between *noema* and *noesis* is illusory since, in the final analysis, one notices a total parallelism between them. It is futile to oppose the unity of the intended object to the multiplicity of the lived experiences which intend it, since each *noetic* moment corresponds only to a definite *noematic* determination. The analysis proceeds, in fact, on the object, and the noetic description contents itself with transposing it onto the plane of the subject. One may conclude that the fresh new philosophy of Husserl takes up once again the traditional critical philosophy; the *absolutist* aspect of phenomenology is only an appearance which evaporates under close scrutiny.

Natorp's critique was the result of a rare act of kindness. The reason for this was not due solely to the friendship that linked him to the author. He agreed, at least in a literal sense, with a major part of the doctrine. Moreover, Husserl had taken Natorp's *Allgemeine Psychologie nach Kritisches Methode* as the basis for his *explication du texte*.

But their points of view remained, in fact, deeply opposed. The Husserlian method was essentially intuitionist, and critical philosophy had its sole *raison d'être* in the negation of the possibility of a concrete knowledge of self. Rickert subjects the notion of the *given* to a radical critique. To take the given as such, as one's point of departure, is to limit it, since everyone has, in fact, only a certain given. It must then be taken in its entire generality, and its definition refers to the concept of the *self* to which it is given. But it could not refer to my own self, since this would be limiting it once again. Hence, the point of departure will be the 'pure' self whose concept alone (construed in a straightforward way) can define the combination of possible *givenness*. The very sphere of the *immediate* is accessible only through a *construction.*[3]

Rickert had not attempted a systematic critique of phenomenology. This task was accomplished by two of his disciples, Kreis and Zocher,[4] who were answered by Fink in his famous article in the *Kantstudien*. It was a matter of demonstrating that phenomenology had been launched in the *Logische Untersuchungen* as a precritical realism, and that in the *Ideen* it could find its philosophical meaning only by adhering to transcendental idealism. Thus, it tended towards critical philosophy without, however, being able to reach it, owing to the persistence of its realist and intuitionist prejudices. Lived immanence, defined as transcendental, remained in fact, psychological; and if Husserl's inspiration bears witness to the value of the critical standpoint, his inability to realize it demonstrates the necessity of doing away with all recourse to intuition.

Phenomenology is, indeed, a philosophical enterprise inasmuch as it poses the problem of objectivity and searches for its solution in a constituting subjectivity. But the Husserlian concept of consciousness was incompatible with the very meaning of a theory of knowledge, inasmuch as Husserl claimed to reach not the pure subject of knowledge to which all objectivity is necessarily referred, but the collection of lived experiences that are given in *intuition*. Intentionality itself would be only a given (among many others) of phenomenological evidence. After having recognized the transcendental problem of the constitution of the object, there was a lapse into precritical and realistic dogmatism. The absolute value accorded to immanent perception is based on the old conception of truth as correspondence with the 'thing itself'. Lived experience (inasmuch as consciousness believes that it is grasping it 'itself') seems to gain an absolute certitude. But this naive confidence accorded to immediate evidence obscures the fundamental critical problem: how is a *given* possible?

These remarks of Kreis are made more explicit by Zocher in a long argument against the 'criterion of immanence'. The fundamental difference which allows for the opposition between consciousness and the world is that "the thing manifests itself in profiles, while lived experience does not do so." Inasmuch as a knowledge 'which manifests itself in profiles' can find itself belied in the course of experience, the status of the world will never be entirely certain. Immanent perception, realizing itself 'without profiles', guarantees its object in an absolute manner. It is clear that such a definition would prohibit our departing from the 'really' (*reell*) lived content of experience. The intentional object would have to be excluded, since it manifests itself precisely 'through its profiles'. But in that case, phenomenological analysis would no longer be transcendental. In order to overcome this difficulty, Husserl asserts, rather suddenly, that the domain of immanence contains not only the 'real' but the 'intentional' as well. In the *Logische Untersuchungen*, the analysis of consciousness remained purely *noetic*. In the *Ideen* the explanation of its constituting signification leads to its extension within the *noematic* plane. But this operation, required by the problematic of knowledge, serves as the precise contradiction of the 'criterion of immanence'.

9. FINK'S REPLY.[1] THE NECESSITY OF A MORE RADICAL EXPLANATION

The thesis of the Kantians rested on the postulate that *transcendental* phenomenology was searching for a resolution of the *critical* problem. It followed that the method employed did not conform to the intended goal. Fink's task was to disencumber the originality of the Husserlian problematic. The theory of knowledge has been merely an occasion for new developments. The intention of the *Ideen* did not consist in wanting to go beyond the 'realism' of the *Logische Untersuchungen* by means of a 'transcendental idealism' (as defined by the critical school). It was a matter of the explication of a task whose meaning already animated the work of 1900.

The fundamental problem of phenomenology does not bear on the determination of the *a priori* conditions of the object. It is none other, in fact, than the very problem which religions and metaphysics attack: the problem of *the origin of the world*. Their answers to this problem, it is true, consisted in laying the foundation for one object upon another object – an absurdity denounced in Kantianism. Yet, the problem remained none the less, legitimate and inevitable. It is important not to evade it, as critical philosophy does by limiting itself to a conceptual analysis of the 'conditions for the possibility of', but to discover a new domain – *outside of the world* – where its solution can be found. Philosophy demands an absolute intelligibility, and intuitionism is only the corollary of such a demand. Critical philosophy, to be sure, also strives to *lay the foundation* of knowledge. But as far as it is concerned, it is a matter of the relation of the existent to the meaning of its being,[2] as a condition of possibility. But the meaning and the existent are both in the world, and the *formal analysis of Kant is, in the final analysis, only an eidetic* (worldly) *analysis*. Phenomenology wants to go back to *the origin of the world*, in order to account for all worldly knowledge in general. A new kind of knowledge which *transcends the world* was necessary for this. For this reason it will be called *transcendental*[3] knowledge.

The phenomenological reduction is the procedure which enables us to transcend the world in order to return to its origin. Obviously, these concepts do not have to be interpreted in a realistic manner. It is not a question of attending to a 'region' which in the spatial sense would be 'outside the world', but of acceding to a kind of knowledge that is inconceivable in worldly life but which specifically reveals the very *truth* of the world. To 'transcend the world' is, for the discovery of transcendental subjectivity, "to maintain the world in the *Universum* as liberated from absolute being."[4] Strictly speaking, it is not a matter of getting out of the world, but simply of going beyond the *narrow-mindedness* of the natural attitude which sees in the world only the *Universum* of the *in-itself*, in order to return to consciousness where being *is constituting itself*.

As a matter of fact, we so far have defined the problem of phenomenology only from the outside. Its clear comprehension – and for this reason critical philosophy could not embrace it – supposes that its fundamental procedure has already been accomplished. The problem of the origin takes on its full signification only for someone who has already abandoned the natural attitude. Worldly problems, such as that of the theory of knowledge,[5] can be the occasion, but not the strictly philosophical starting point. Hence the necessary imperfection of the exposition of the *Ideen*, where one started with the world in order to transcend it. Thus, transcendental notions are put forth initially in the language of the natural attitude: the explication of their meaning provokes intolerable contradictions, which motivate the reduction. Following this, the exposition returns to the same subject, but in a new language. Whence the surprise of the reader who believes it to be an incoherence of thought.

The first thematization of the transcendental act – the act of positing the world – itself takes place in the world. A human being is the one who perceives objects. But analysis shows that the whole meaning of objectivity resides in this very perception. The world exists only through me who am in it. This absurdity provokes an 'inordinate shock' which incites us to effect the *epoché*. The exclusion of the world as an *in itself*, reveals it to us as *transcendental validity* in the becoming of a constituting subjectivity. This subjectivity is that very same one that functioned spontaneously in the natural attitude and which now is found disclosed in its *truth*. The guise of psychologism comes from the fact that the reader continues to interpret as worldly that which is revealed precisely as transcendental, and does not understand that transcendental life was already immanent in natural life. Consciousness, in being thematized, continues to live its constituting activity. The world does not lose its being *in itself* to become a simple *for us*: all it does is explicate the authentic meaning that it has always had in the naive state of existence which characterizes the natural attitude.

Thus, phenomenology can be confused with neither a critical philosophy nor a psychological theory of knowledge. It does not seek to reascend from the constituted objects to the constituting forms but rather to reach the concrete source of ultimate intelligibility. But it is not any longer a question of psychological immanence, as opposed to the 'external world'. The 'criterion of immanence' whose inadequacy Zocher has demonstrated, was understood correctly as only a provisional characteristic, inevitable in the natural attitude which served as a starting point. It was necessary to begin with the paradoxical task of defining the being of the transcendental within the horizon of the world. The immanent, then, is opposed to the transcendent and appears as a particular *region*. But the actually thematized field encompasses both: their unity is precisely what constitutes the truth of consciousness and of the world.

No one was better qualified than Fink to interpret Husserl's thought. Daily contact with him enabled Fink to fathom the most subtle nuances of Husserl's thought. But perhaps too great a familiarity has prevented him from recognizing difficulties which deter those who still live in the naiveté of natural existence. The *intentions* of phenomenology are explicated in a definitive manner, and the article which we have just analyzed can serve as a *guiding thread*: it would be difficult to find an actual *demonstration* in it. We learn that it is not a matter of search for the *a priori forms* of the object but for the *origin of the world* – with the major efforts of religions and metaphysics striving towards the solution to this ultimate problem. But if we ask them to be more precise, we are told that the reduction provides its meaning only to those who have already performed it. The transcendental phenomenologist, says Fink at the end of his article, returns to the world in order to share his discoveries with the dogmatic philosopher: but he cannot convince him . . . .

Let us pose the problem in precise terms. Phenomenology discloses the true *being* of the existent as *being-constituted* in transcendental subjectivity. If we look back to the theory of constitution, we notice that it is dominated entirely by the structure of the *Dinglichkeit*. Animals, men and 'cultural objects' are constituted upon a matrix of materiality, which serves as their ultimate and universal support. 'Idealities' depend, in a sense, on sensible objects: even for logical forms, 'their kernels are relevant'. We cannot expatiate here on the extraordinarily complicated details of the *Dingkonstitution* which was the constant object of concern for the author. It will be enough to recall its general principle.

The thing is given to us in a series of silhouettes or profiles (*Abschattungen*) which refer to each other within the 'horizons' determined by the meaning of the intentional act. This table which I perceive, 'profiles itself' to me under certain aspects, with each calling forth others in an infinite unfolding. Thus, I know that if I were to take a different position, I would see it differently in a way which is unforeseeable in its concrete particularity, but relatively determined beforehand by the implied meaning of my actual perception inasmuch as it is a perception *of* this table. These 'horizons' comprise a structure which is characteristic of the existent in question: thus, spatio-temporality designates a fundamental group of *a priori* rules for the syntheses which are unfolded in the experience of the 'thing'. The course of perception consists in an enrichment and continual corrections within a definite form. It is these syntheses which 'constitute' the intended object as an ideal unity. It is to be noted that it is transcendent, not immanent. No series of *Abschattungen* will succeed in exhausting it. In itself, however, it is only the ideal focus to which this unfolding refers. All its *being* consists in *being constituted*.

Such is the theory which would serve at one and the same time as the foundation and model for the constitution of the world in general. We do not yet see in what way it is actually distinguished from Kantian conceptions. The moment of objectivity is defined in terms of a certain number of *a priori* relations within immanence which characterize the perception of the thing. These relations, moreover, are not detached from an eidetic analysis of the lived but are taken originally from the object itself. As Natorp clearly saw, *noetic* description does nothing more than transpose to the plane of the subject the results of the *noematic* analysis. How then can we consider the pretension of directly thematizing concrete and temporal consciousness, qualified for the occasion as 'transcendental consciousness', as anything other than a simple persistence of psychologistic prejudices?

Phenomenology claims to explicate the authentic meaning of the status of the world. But the constitution which it presents to us does not seem at this point to satisfy these requirements. It is quite clear that thought intends in the object something other than the ideal correlate of a synthesis of *Abschattungen*. The solution of the *Ideen* ended by defining for every kind of reality, an "equivalent in consciousness."[6] This was a falling back into the absurdity of the philosophy of the *as if*. Such an *idealization* contradicted the absolutist pretensions of Husserlian thought. The definition of the objectivity of the object by the *a priori* forms of synthesis was legitimate in Kant, who wanted only to consider it as a 'phenomenon' and not as a 'thing-in-itself': morality, works of art, and forms of life were practical ways of reaching an absolute – the access to which was forbidden to theoretical speculation. These distinctions and reservations are unknown to Husserl, whose mode of constituting claims to bear upon the *thing itself*, It is a matter of thematizing the *origin of the world* and of responding to the *ultimate questions*: this is all the more irksome to the reader who finds only a system of *Abschattungen*. But perhaps there has been a misunderstanding. Despite a certain clumsiness in its expression, the originality of phenomenology is unquestionable: it imposes itself in the very *style* of concrete analyses. It thus is appropriate for us to search anew on the basis of other examples for the actual meaning of constitutive investigations.

10. THE NOTION OF CONSTITUTION. THE SIGNIFICATION OF TRANSCENDENTAL IDEALISM

We shall simply follow the exposition of the constitution. Perhaps difficulties will arise from the *thing itself*, which will oblige us to make our concepts more precise.

The existence of the 'thing' (*Ding*) was defined as the ideal correlate of the coherence of sensible syntheses. But if we push reflection further, we perceive that it can, at first, refer only to *my* syntheses. Yet, the objectivity of the object refers to experiences other than mine. It is defined as precisely the correlate of their universal agreement. But how can I speak of *other* consciousnesses, since only mine is *given* to me, in the strict sense of this term. The reduction, by eliminating all transcendence, also has set aside that of *the other*. The primordial phenomenological domain can be nothing other than the collection of lived experiences which constitute *my* own life, in the unity of *my* stream of consciousness. If other 'me's' exist in addition to me myself, they evidently can have meaning for me only as 'in myself'. In other words, I must *constitute* others by starting with myself, and this constitution is required for the very constitution of the 'thing'.

How does the other self 'appear' to *my* own consciousness? This meaning has a particularity which deserves to attract our attention. He is not and can never be, strictly speaking, *given* to me. It belongs to the essence of the other self to be absolutely 'other', inasmuch as he is precisely a 'self'. His transcendence does not simply refer, as that of the 'thing', to an infinite series of silhouettes (*Abschattungen*). The other self is a 'self' *as 'I myself'* am; he possesses, outside of me, the mode of existence that I myself possess – *transcendental existence* as *absolute existence*. Thus, the problem of 'constitution' is posed in paradoxical terms.

> How can my Ego, within its peculiar ownness, be capable of constituting in some manner 'the other', since rightly being a stranger to him, that is to say, how confer on him an existential meaning which places him beyond the concrete content of 'me myself' in its concreteness as constituting him? [1]

The other self, inasmuch as it is a 'self', will never be 'in me'; nevertheless, I can never comprehend it if it is not in me, and the constitution of the other is necessarily accomplished in me myself.

The task is pursued in the fifth *Cartesian Meditation*, in details that need not detain us. It is enough to recall the essential points. The procedure is performed in two parts. A first constitution, by means of syntheses of *Abschattungen* inside of my own self, *solus ipse*, concludes by posing a first 'primordial' nature. This nature carries with it no other meaning than that which is related strictly to me myself. In this primordial world, I find my body, characterized as 'mine' and also others, outside of me, to whom I do not accord, as yet, any 'subjective' existence. But among them, there are some who by analogy, by virtue of a certain resemblance to my own body, take on the meaning: 'body of another'. Of course, it is not a matter of reasoning: the foreign body enters the field of perception 'coupled' with mine, and this 'coupling' enables a transfer of signification, thanks to which I see, before me, *another myself*.

This brief exposition does not pretend to reproduce the extraordinary subtlety of the analysis. It is enough, nevertheless, to point out its major fault. Without entering into the details of the description, we see rather clearly that they proceed from a method which, to adhere to concepts that already were utilized, does not actually go beyond the constitution of the *Dinglichkeit*. I perceive this physical body, in the sense of 'human body'. But how can I be assured that it really is so? By 'concordant verifications'. It is true that the author does not say 'by *Abschattungen*'. The self of the other is perceived in the monadic unity of his lived experiences, and not as a *thing* which manifests itself in profiles. But this puts him in a poor position to explain the difference since once more it is a matter only of a system of syntheses where the object, in the final analysis, is nothing more than the *correlate* of the unfolding of *my* lived experiences inasmuch as it is realized in a coherent manner: it is the *ideal unity* intended in my own sphere.

However, the explication of the meaning of the other at the beginning of the *Meditation* had revealed an exigency which should have motivated going beyond this method. 'The other self' is not simply a truth intended *in me*, a transcendence to which I would accord only an *intentional* existence as opposed to my own *real* existence. It truly concerns an 'other self' who, as such, possesses the same existence as 'I myself'; i.e., the absolute existence of consciousness. It was manifestly contradictory to want to 'constitute' an 'object' of this kind with the procedure of concordant syntheses. This implies a depreciation of the notion of reality, which makes the world, as the correlate of consciousness, become a dream – albeit a 'well-ordered' one. This thesis could be admitted, if necessary, in the case of the 'thing', whose absolute existence can easily be contested. But the problem of the 'other' raises a difficulty which requires an original solution. Never will anyone be made to believe that his *being* consists in *being constituted* in the synthesis of my experiences. If the word 'constitution' still ought to be used, it would have to be only in a radically new sense. It is especially essential to eliminate every kind of *idealism* from it in the sense that idealism excludes the status of an *in itself*. The other is not a simple unity intended by me, but it possesses the fullness of existence, in the strongest sense of this term. *Transcendental ideality* should not correspond simply to empirical *reality* but also to *absolute reality*.

Such an exigency demanded new methods. We find this to be the case in some of the unpublished texts. Thus, the manuscript of *Ideen II* strongly opposes the constitution of persons to that of things. Each person, says the author, is *not* a 'unity of appearances' (*Erscheinungseinheit*) but a 'unity of absolute manifestation' (*Einheit absoluter Bekundung*). Persons are 'constituted in an absolute manner'; things, in a 'relative manner'.[2] Thus, the absolute existence of persons is comprehended in an 'absolute manifestation' in opposition to the 'unfolding of silhouettes' (*Abschattende Darstellung*) which characterizes the perception of bodies. These two modes of experience create the opposition of two worlds: one, relative – the *world of nature*; the other, absolute – the *world of spirit*. "The reign of nature is the reign of real 'phenomenal' unities; that is, they are constituted in or by an 'unfolding'

(*Darstellung*); the reign of the spirit is the reign of realities given by means of an *absolute manifestation*."[3]

We can now understand the sense of the misunderstanding which placed the disciples in opposition to the scholar after the first volume of the *Ideen* had been published and which proved fatal to the destinies of the phenomenological school. The privileged example of the *Dingkonstitution* had understood transcendental idealism as the negation of the absolute existence of the object. Such a negation appeared to be contrary to the deepest intentions of the doctrine. It related phenomenology to a rather incoherent critical philosophy, since the consciousness to which it returned was a temporal consciousness. But for Husserl, the identity of *being* and *constituted being* could not, by itself, entail any 'idealization' of the real. It was simply a matter of grasping the meaning of the existent inasmuch as it is revealed in concrete subjectivity. This takes the name 'transcendental', seeing that the *being* of the object can be none other than this very *meaning*. If the 'thing' is reduced to being only a simple ideal correlate of syntheses of *Abschattungen*, the reason is that it does not seem to possess (at first sight, anyway) any other mode of existence. But the 'relativity' of its being can hardly be extended to the general notion of *constituted being*. In the domain of absolute existence – the spiritual existence of persons – phenomenological idealism reveals the fullness of its meaning. The person is not a unity of appearances, but a unity of absolute manifestation. There could no longer be any question of a 'reduction' of *real* being to *intentional* being; yet, intentional being is here fully real, since such a reality belongs precisely to its very meaning. There is no more in the 'other' than what I intend in him. But I truly intend an absolute existence, and this existence should be granted to the object of my intention. To say that his *being* consists in *being constituted* is to say simply that a person who in principle would be inaccessible to me would not exist. The necessity of constituting the spiritual world, starting with *my* consciousness, entails, beyond a doubt, a sort of 'relativity'. But this 'relativity' is of an entirely different order from that of the 'thing'. There is no question of denying the reality of the 'other', but simply of asserting the *a priori possibility* of a *communication of consciousnesses*. For it belongs to the essence of every person *other than me myself* to be capable of being given *in me*, since his very meaning as a 'self' can only arise from this perception wherein I grasp 'another me-myself'. But such an experience implies precisely an absolute alteriority.

The author could claim, and rightly so, to be faithful to his original intentions. The notion of *constituted-being* actually realized for him the fullness of the notion of *being*. Nevertheless, it remained true that the exposition, limited as it was to the theory of the 'thing' (to which an exemplary value was given) made credible in an irresistible manner an idealism of Kantian inspiration. Only a coherent constitution of the spiritual world could have done away with this confusion. The task was never accomplished, but, perhaps, an examination of this *failure* will make us better understand the meaning of the project.

11. THE CONSTITUTION OF THE WORLD OF THE SPIRIT

The first volume of the *Ideen* gave an explanation of the object and the method of transcendental phenomenology and also gave an example for the case of the 'thing'. The achievement of the whole project was reserved for the second volume, which was never ready for publication. We can surmise that this was the result of certain faults in the basic definitions. In the preceding pages we have tried to rediscover the authentic meaning of transcendental idealism, as it forces itself upon us whenever we depart from analyses of the 'thing'. It is necessary to corroborate this interpretation with the *results* of actual investigations.

Volume II of the *Ideen* is composed of three parts corresponding to three fundamental regions: *material nature, animal nature*, and *the world of the spirit*. The first takes up once more in more detail, the constitution of the 'thing', of which we have already spoken in regard to the first volume. Animal life is constituted on matter, which is the fundamental level. It deals not only with animals, properly so called, but also with man as a being of nature. The soul, which is the subject matter of psychology, is a level 'founded' upon the body, and because of this relation of 'foundation',[1] is inserted in the spatio-temporal world. It is 'perceived' in the *Einfühlung*, or perception of the other, on the 'foundation' of the presence of his body. But, as the author remarks at the beginning of the third part, each level constitutes in itself a unity which can always be studied 'apart'. If we take human life, abstracting it from all connections with nature, we are taking a new attitude – the *personalistic attitude*: it is that very one which characterizes the existence that we lead in society.[2]

The text can give rise to confusion. One would be tempted to believe that the person as such is an abstraction. But this is not the case at all: the consideration of persons as such is presented as a general attitude which defines, correlatively, a world that is self-sufficient – the *world of the spirit*. The latter encompasses not only persons, but also things, inasmuch as they are *for* persons: not as physical realities in themselves, but as intentional objects of human consciousness, whose activity they motivate. The relation of man to nature is no longer constituted by causal relations but by *phenomenological relations*. Things are not defined *for me* in terms of their physical properties but in terms of their experienced aspects, by their predicates of value and action. They *are* for me only inasmuch as I *perceive* them with the *meaning* that they have for my actual life, and all their *spiritual being* consists in this meaning. The *naturalistic attitude* to which the world of nature corresponds (the subject matter of the sciences of nature, including psychology) was a *theoretical attitude:*[3] it grasps the real as an *in-itself* independent of the subject which knows it. The *personalist* attitude which takes the world as the world of the spirit and as object of the sciences of the spirit, is characterized as a *practical* attitude:[4] the real has meaning only in its relations with the concrete subject in his actual life. *Being* is defined as *being for me*, not in the speculative sense but in the sense that is *experienced in practice*.

Now persons no longer appear as psychical realities *founded* on material bodies: this would be a return to the naturalistic attitude. They arise as *subjects* that I perceive in the experience of comprehending. The other self, *like me*, is a subject of a world which exists only *through him* and which exists only to the extent that it is *for him*. And I understand at the same time that this world is the same as mine, that *we live in the same world*. Such a signification is made possible precisely by the practical horizon by which we have defined the personalistic attitude. Though the relation of pure knowledge depreciates the reality of the object by reducing it to the universality and necessity of the series of its appearances in consciousness, *lived practice* reveals the real in its absolute existence, as a *subjectivity other than mine*. I cannot doubt the existence of the *other* in the experience of relations in which he acts upon me, at the same time that I act upon him. In these reciprocal actions, each acts by virtue of being a *subject*, and the *object* upon which he acts is equally a *subject*. It is not at all a matter of a causal action of mechanical displacement, but of an *influence* whereby each person, in presenting himself to the other as a person, *motivates* the other to perform an activity. This motivation operates precisely by means of comprehension itself: *persons determine one another by making themselves understood.*[5]

We see now the origin of the obscurity of Husserlian concepts. 'Transcendental idealism' had two entirely different goals. On the one hand, the explication of the *Dinglichkeit* as *ideal correlate* enabled one to bypass the concept of a nature *in itself*, in the sense in which it would prevent the comprehension of a strictly human existence whereby being *is* only inasmuch as it is *for me*. But, on the other hand, it was necessary to disengage the meaning of the world of persons, inasmuch as its *being for me* implies *existence* itself, in the full sense of the term, since I perceive it precisely as a collection of *subjects*; i.e., of other '*me's*'. The same concept of 'constitution' now refers to the positing and recognition of an *absolute reality*.

This duality of signification was never clearly explicated for the author himself: in our opinion, therein lies the reason for the incompleteness of the *Ideen*, and more generally for the Husserlian endeavor itself. The theory of the world of spirit was inserted into the general project of the *Weltkonstitution*, interpreted according to the concept of *Dingkonstitution*. But the originality of the domain should have produced a new framework wherein the very notion of 'thing' would have had an entirely new signification. It was quite clear that, whatever might be the specific feature of the 'personalist attitude', it could not do without corporeal existence. The understanding of persons ineluctably implies "the perception of the body (inasmuch as it constitutes the founding corporeity) as the foundation for the perception of spiritual meaning."[6] But with materiality being recognized as having only an existence that was purely phenomenal, the absolute reality of the world of spirit becomes thereby totally unthinkable: how can real existence be founded on a mere 'ideal correlate' of subjective syntheses?

The author was not unaware of this difficulty. But, strangely enough, it is precisely because of its very ideality that he claimed to integrate the *Dinglichkeit* into the world of the spirit. He explains that, in terms of his personalist point of view, the body is defined in a purely 'phenomenological' manner, by means of the synthesis of appearing modes.[7] This, most assuredly, was falling into complete absurdity. The constitution by synthesis of *Abschattungen* had meaning only within the framework of the theoretical attitude which intends, in fact, only an ideal unity (*focus imaginarius*) of subjective representations. In the world of spirit, perceived in the practical attitude, existence, *including the existence of things*, can be taken only in its absolute sense: for, as is quite clear, practice is possible exclusively on realities. Whence, the upholding of the constitution by means of modes of appearance envelops a contradiction, confirmed by a total idealization – the world of spirit being founded on *Dinglichkeit.*[8] Since all experience implies the moment of the *thing*, the dialectic which rises from matter to spirit can terminate in a reality solely, on the condition that the starting point is itself a reality. The reality of the world of persons being revealed as incontestable, it follows that its material foundation in no way admits of *idealization. The Husserlian doctrine failed to constitute the thing without reducing it to an equivalent in consciousness.*

To be sure the accomplishment of such a task would imply the complete bankruptcy of transcendental idealism: the reality of *Dinglichkeit* can only be recognized in a new kind of naturalism. But before looking at this problem any further, it is necessary to sharpen the concept of constitution by studying the consciousness where the object is constituted.

12. THE NOTION OF OBJECT. PERCEPTION AND JUDGMENT

Phenomenological analysis as intentional analysis finds its culminating point in the theory of evidence. Since, in itself, consciousness implies not only its 'really' experienced content but also the intended object as such, the meaning of reality will be disengaged from the structure of the act wherein the object is not *simply intended* in a more or less distant way, but actually and fully *present in its own right*: the description of lived evidence will make possible, at the same time, the definition of the objectivity of the object.

It is clear that such a proposition implied a new definition of fundamental notions. According to classical perspectives, it led directly to psychologism: though the proposition $1 + 2 = 2 + 1$ could very well be intended by an obscure consciousness, the analysis of evidence wherein it was given will not define its truth-meaning (independent of the subjective course of our representations) but simply the psychological conditions by which at certain privileged moments, we become fully cognizant of it. The *objectivity* of the object is precisely that by which the *subjectivity* of our lived temporal experiences is denied and bypassed: hence, as Husserl claims, to grant to the description of concrete consciousness the value of a determination of the conditions of the object is, from the point of view of traditional philosophy, to fall into a *transcendental psychologism*. But the originality of phenomenology rested precisely on a radically new conception of the relations of knowledge and (even more generally) of the relation of thought to being.

The definition of the notion of object starts with common-sense considerations. When we speak, we speak *of some thing*. Our propositions, enunciated as true, can be so only about *the very things* of which they are statements. The *object*, in the most general sense, is the 'subject' of discourse: 'that of which we speak' as opposed to 'what one says about it'. But this gives us only a formal definition. What must one understand by 'that of which we speak? We fall back here into a classical difficulty: what will substance be outside of its predicates, since these express precisely what it *is*? The whole truth of reality seems to exhaust itself in its intelligible content, and intuition would represent only a provisional moment of knowledge or (to put it more precisely) it would be brought back upon itself in its positive content – the concept. But then the problem of knowledge is placed immediately on the plane of universality, since the conditions for the possibility of the object are determined not by the lived experience of immediate evidence but by the intelligible structures of predication. But how can man, being temporal and finite, raise himself to a *truth*, and how, moreover, can this truth, defined as it is by the universality and necessity of the concept, be attributed legitimately to the *reality* of things given in experience? This was the double *aporia* to which traditional thought could answer with only verbal solutions, all amounting to a more or less modified form of the mythology of *participation*. Husserl's merit was in having posed the problem on the concrete plane, thereby allowing for methodical progress in its comprehension. 'That of which we speak', the *upokeimenon*, is not a simple indeterminate and empty substrate: it is the object *itself*, just as it presents itself in the antepredicative evidence of perception. I say of this table that it is small. The asserted content, 'the table is small', refers to that of which it is asserted, namely the table itself that is given in experience. The *perception* of the table, and this alone, can verify the *judgment* that it is small.

It is necessary to be more precise here. There could hardly be any question of transforming the judgment into a copy of perception: this would be returning to the absurdities of a long abandoned empiricism. No one will any longer dream of denying that the perceptions of civilized man imply an infinity of predicative significations that are more or less sedimented. But perceptive experience contains a solid core anterior to the judgment, which cannot be reduced to a 'rhapsody of sensations' but constitutes precisely the first sense of objectivity. In a more general manner, the object of perception, even in its intelligible determinations inherited from anterior predications, is defined as an *actuality* which places it irreducibly in opposition to the *ideality* of the object of judgment. Thus, perception possesses an original structure upon which the judgment is constituted. The concept, far from dominating the synthesis of intuition, must always be referred to the intuition; for, inasmuch as a predicate can be enunciated as a predicate of *this very thing*, and of it alone of which it is an assertion, its *truth* will refer necessarily to a *verification* of the foundation of the *perception of the thing* where the judgment itself is constituted as self-evident.

Whatever may be the difficulties that the reader accustomed to classical texts could not help but encounter, these difficulties should not make him misjudge the faithfulness of the description. The notions are taken in their original sense, just as they stand forth before the gaze of the phenomenologist. The theory of perception, as anterior to judgment and founding its truth-value, has been set forth consistently, taken up again, and developed from the time of the *Logische Untersuchungen* up to the writings of the latest period. It goes beyond the framework of problems proper to logic and the theory of knowledge and circumscribes the very phenomenological inspiration summarized in the well known motto: "Return to the things themselves." *Zu den Sachen selbst*! These 'things' must be understood as that which is given in a 'perception' (in the broadest sense of the term) prior to the concept – as a *primordial intuition* to which we must always return in order to verify and correct our conceptual constructions.

To characterize the originality of perception in contrast to that of judgment, we shall concern ourselves with the structure of the contents which each intends. The intentional object posited in a judgment, the *state of affairs* (*Sachverhalt*) is an ideal signification, as exemplified by, 'the table is small'. By stating this proposition as true, we assert it to be valid for everyone and always. Yet, even though this is an example of a judgment *for an occasion*, whose verification is possible only within the horizon of a historically determined situation, still, its meaning constitutes a pure ideality which refers to a *de jure* universality. The table may disappear or be modified. But the fact that it is small at this moment constitutes a *truth*, and it belongs to the essence of truth to be universally and eternally valid. In other words, the properly intended object in the judgment, the *state of affairs*, is an *ideal object* which is not found in either space or time, since it is the *same* for everyone and always.

It is an altogether different matter regarding the object given in perception. I perceive the table as existing contemporaneously with *my* existence. After its destruction I shall still be able to *judge* it, but I shall no longer be able to *perceive* it. A true judgment can always, in principle, be enunciated, no matter what the time or who the person. It belongs to the essence of perception to be realized only in an actual subject. The actuality of the intuitive datum is essential to the perceiving consciousness inasmuch as it can take hold of only a certain number of objects at each moment – the very same ones which stand out within his field of presence; whereas, the total system of conceptual truths is, *de jure*, accessible to the subject of predicative knowledge at every moment.

The object of perception is that particular object possessing a unique existence. To be sure, intuitive consciousness is not reduced to the present instant as such. The past event is retained in memory, while the future is the object of anticipation. But it always deals with *this* particular existence which, once past, can no longer be perceived but only remembered. Memory *itself* graps it precisely as past. This absoluteness of temporal existence is, moreover, not an *in-itself*, which would impose itself from the outside; it is only the correlate of the temporality of *my* actual consciousness. For me to remember a past event is for me to return to my perception of that event. The past of the world has meaning only through *my* past which refers to *my* present. My past is what I *still* am *at the present moment*: the past exists as past only in reference to the present, where it is inserted in the form of the *still yet* (*encore*). The reality of the object is defined by my *actual* consciousness.[1]

One could think that here it is a matter of a condition that is simply 'psychological'; according to traditional prejudices, this would be closing one's eyes to *the thing itself*. It belongs to the essence of the *real* object to be capable of being given to only an actual consciousness. The *predicative* content, on the other hand, is by definition accessible as such to the universality of all possible subjects. The 'state of affairs' intended in the judgment is always an ideal object that is identical, in principle, for all, and at all times and places. But the real object is essentially particular: that very thing which I perceive and whose particularity refers, in the final analysis, to *my* actual particularity.

It is not a question, of course, of that unseizable 'shock' to which one is accustomed to reduce the proper content of temporal consciousness. The object of perception is not a 'rhapsody of sensations' but a well defined unity, although this object, in itself, does not have, as yet, any conceptual structure. Its 'configurations' refer only to antepredicative activities which are diversified, depending on the experienced situation of actual subjects. Permanence is presented in a concrete form and implies a variety of irreducible aspects: the past is given to me in memory solely as past, and in anticipation the future as future. A scene which I have not seen can be related to me, but it is given to me exclusively in the narrative which itself refers to the perception of witnesses. The concern here is not an *abstract universality* but the *totality* of the actual consciousnesses inasmuch as they all constitute the same world.[2]

On the contrary, the proper object of the judgment – the state of affairs – is essentially the same for all. It refers *a priori* to the abstract universality of possible consciousnesses to which it is presented always with the same content. A proposition, by definition, always retains the same meaning for everyone; whereas the object of perception immediately assumes the meaning of 'past' and is buried in memory. The real object, which is constituted essentially in the actuality of a particular consciousness, is given in a different manner to different subjects. The 'ideal' object of judgment does not vary according to the situation of the self which posits it. This peculiarity has inspired classical rationalism with the conception of an eternal and universal subject that is hypostasized in God, the fount of all truth. The real world is eliminated to the advantage of a system of propositions which can be affirmed about it. Sensible knowledge becomes incomplete intellectual knowledge – consequences without their premises. Inasmuch as every truth is eternal by definition, that alone exists which is sustained by the eternal spirit.

This was to ignore the authentic meaning of ideal objects. Universality is the '*pretension*' of the act which posits these objects; it is universality which explicates the first moment of analysis. But further reflection discloses a hidden intentionality which grounds its meaning. That this table is small, constitutes an intelligible content which, seeing that it remains 'the same for all', is postulated as outside of space and time. But, evidently, it has a meaning only if it refers to the table itself, given in an actual perception. Its whole *truth* depends on the possibility of a *verification*. What we are here dealing with is not simply an empirical circumstance which, in a contingent manner, would condition the access of human consciousness to the absolute domain of eternal truths. The whole meaning of predicative truth consists in being true *of* the object of which it is asserted. This meaning is disclosed in the act by which it arises with evidence *regarding* (*sur*) the intuition of the thing itself. The whole *being* of idealities consists in *being constituted* in such a disclosure.

These considerations bring us to the definition of evidence in a case in which the ideal object is 'itself' given, in opposition to the empty intentionalities when it is 'simply represented'. We shall say that the object of judgment is given in a 'categorial perception' (*Kategorein*: to predicate).[3] The term must not make us fear a return to empiricism. It concerns a perception which grasps not a real object but rather an ideal object: it is precisely the very act of evidential judgment. Nevertheless, it is a *perception* since *the thing itself* is originally given in it.

The preceding analyses enable us to define categorial perception as *founded* perception, inasmuch as it can be realized only on the foundation of sense perception. Every proposition stated on the simple plane of discourse – 'simply represented' – ineluctably refers to such a 'perception' which alone can give it the fullness of its meaning. Intellectual knowledge is not a knowledge 'apart'; it always presupposes sense knowledge as its foundation.

From the above, we can now understand the profound reason for the extension of the notion of 'perception'. It seemed at first to rest on some sort of verbal convention. We understand now that there is a more than formal analogy between categorial and sense perception. Sense perception, as we have said, grasps a reality whose meaning irreducibly refers to the *actuality* of the constituting consciousness. Ideal objects seemed, at first, to exist in an eternal subject. But phenomenological explanation allows for the correction of this error. The constituting consciousness of idealities is a 'founded' consciousness. Inasmuch as its 'foundation' rests upon a sensible intuition – temporal – *the ego (le moi) as the bearer of predicative truths will itself be a temporal self*. The matter at hand here is not an empirical circumstance but a transcendental condition of the objectivity of the object.

The idealities posited in the judgment, in fact, take all their meaning from the act which constitutes them in the presence of the sensible. The ideal object is none other than *that very same* which is presented to us in the 'perspective' evidence of a *founded* judgment. To interpret its eternity as that of an *in itself* is to fall into intellectual realism. In fact, the eternity can be only the *meaning* of the actual intentional act inasmuch as it *constitutes* an ideal object, as independent of the particularity of actual consciousnesses. But the negation envelops that of which it is the negation. *It is a temporal subject which posits an object as transcending its own temporality, and the whole meaning of eternity consists in this experienced negation*. The ideality of the categorial object does not imply a belonging to a world 'apart' from ideas: it is a simple matter of asserting that it can be posited for everyone and always. Universality is but the correlate of this very act: *atemporality* has no other meaning than *omnitemporality* which is itself but a *mode of temporality*.[4]

Thus, the extension of the notion of object – and, correlatively, of perception – is not reducible to a mere convention of words. It corresponds to the meaning of the 'thing', inasmuch as all knowledge refers to a particular subject, whose actuality is not a simple psychological given but a condition of objectivity. The constituting self is an actual self because *every object, whether real or ideal, finds the ontological meaning of its being in the temporality of its transcendental life*. It is in perception as a lived act that the *thing itself* is constituted.

There is a profound tendency of the human spirit at the present stage of its development, which propels it to realize truth as an *in itself*, absolutely independent of the subject which posits it. This is not a simple illusion resulting from the naiveté of the natural attitude. It belongs to the essence of truth to transcend consciousness, and this is precisely the meaning of the constituting act. But the error of classical rationalism was to take up once more on the philosophical plane what, manifestly, had meaning only for spontaneous life. The scientist, quite legitimately, posits his theory as an absolute; for he started, in fact, from sense perception, and his judgment is founded on that very same basis from which he abstracted it. The relation of *founded* to *founding* is defined precisely as a negation of that kind. The scientist's dogmatism does not constitute any danger, since his 'going beyond' the sensible is carried out by him with all its lived meaning: different experiences will lead him to devise different theories. The philosopher places himself on the conceptual plane from the outset, and the naiveté of rationalism consisted in hypostasizing the idea of truth into a total system of a supposedly completed science. This involved forgetting the original conditions of the scientific effort.

The merit of Kantianism was in having reminded us of the actual meaning of predicative operations – their transcendental character. The simple concept enables us to 'think', not to 'know' – knowledge being possible only within experience. It would seem, then, that the critique would have to begin by defining in itself the world of naive perception, since it necessarily grounds the value of all predication. In fact, the *Transcendental Analytic* was preceded by an *Aesthetic* which gave us the conditions for a possible intuition – space and time as *a priori* forms of sensibility. But from the very beginning of the *Analytic* the distinction between intuition and concept raised the problem of their necessary agreement:

> Appearances might very well be so constituted that the understanding should not find them to be in accordance with the conditions of its unity . . . . But since intuition stands in no need whatsoever of the functions of thought, appearances would none the less present objects to our intuition.[5*]

Thus, had the problem been posed in a lucid manner, it could have received an authentic solution by means of the analysis of the act which, constituting the concept upon intuition, explicates, on the universal plane, the *truth* of the life-world without, for all that losing its originality, since the whole *truth* of the system of science rests on the possibility of a *verification* in experience. But a procedure of this kind supposed a radical reform of the notion of object. Realities given in perception carried, in their irreducible particularity, an essential relation with the actual act which perceives them. It is not a question of a relativity that is explicable in terms of the insertion of psychological subjectivity into the world of the object. The real meaning of reality implies the actuality of perceiving consciousness as a *constituting* consciousness.

But, in spite of the inspired intuition that brought Kant to condition the *Analytic* by the *Aesthetic*, he remained attached to the classical notion of objectivity. The object, conceived on the model of the object of judgment, is what is posited as essentially independent of the particularity of the actual subject. And since phenomena do not exist in themselves but are merely 'representations', objectivity can be defined solely in terms of the necessity of the synthetic unity which, being performed according to universal rules, escapes the '*subjectivity*' of particular consciousnesses. Hence, perception is itself grounded in the concept, under pain of having no *object*, since "the object ... can be nothing other than the formal unity of consciousness in the synthesis of the manifold of representations." The concept designates precisely the set of rules by which the representation acquires an *objective* value. "Thus we think a triangle, as an object, in that we are conscious of the combination of three straight lines according to a rule by which such an intuition can always be represented."[6 *]

Hence, the agreement of the sensibility with the understanding might be deduced with analytical necessity. Inasmuch as the concept defines the objectivity of the very object of perception and the categories are the 'concepts of an object in general," it follows clearly that all perception, as perception of an object, must conform to the categories. It would be difficult not to admit the logical rigor of such a procedure. Its very elegance should have made its legitimacy suspect. How might one obtain a manifestly *synthetic* result by means of an *analytical* method? The analysis of the content of the concept of the object has value only if that very concept has been adequately understood from the very beginning. That this is not the case will be demonstrated conclusively by the immediate consequences of the doctrine.

We will not belabor the point regarding the inextricable difficulties raised by the relation of the *Aesthetic* to the *Analytic*, seeing that space and time (after having been presented as independent form given in pure intuition) suddenly appear as the result of determinations brought about according to the unity of the understanding. The eccentric distinction between the 'form of intuition' and the 'formal intuition' in the famous note to Section 26 of the *Deduction* bears witness to the incoherence of the process.

Space, represented as *object* (as is really necessary in geometry) contains more than a mere form of intuition, it also contains *combinations* of the manifold, given according

> to the form of sensibility, in an *intuitive* representation, so that the *form of intuition* gives only a manifold, the *formal intuition* gives unity of representation.[7*]

That this unity refers to the understanding is what results from the most elementary Kantian definitions, and Kant himself specifies "that it presupposes a synthesis which does not belong to the senses."[8*] But this involves suppressing the autonomy of the *Aesthetic* and absorbing the whole meaning of objectivity into the *logical* conditions of Knowledge. But the author is eager to add with an astonishing mode of reasoning that " . . . since by its means (in that the understanding determines the sensibility) space and time are first *given* as intuitions, the unity of this *a priori* intuition belongs to space and time and not to the concept of understanding."[9*] As if transcendental analysis, by right, did not have to refer each moment of knowledge to its origin! Since the formal intuitions of space and time are themselves the product of a synthesis where "the understanding determines the sensibility", there is no doubt that the *Transcendental Aesthetic* should only form a chapter of the *Analytic*, inasmuch as the categories alone define the structure of objectivity. Thus the abstract development of the system compromised the profound inspiration of the fruitfulness of the *Critique* – the feeling of the irreducible originality of the *intuition* as opposed to the logical conditions of *thought*. But it is not part of our intention to examine the *Critique* in detail. It will suffice to tackle *the thing itself* – i.e., to understand precisely the notion of object.

The concept of object (once having conceded Kant's definition of object) will be *ideal* and not *sensible*. No doubt, the retort will be made that we have only to give rules according to which the object is determined in intuition. But the rigor of phenomenological analysis is precisely what is needed to dissipate such confusion.

To make this point clear, let us consider a particular case – e.g., that of causality. A succession, asserts Kant, possesses an *objective* value on the condition that it is produced *according to rules*; otherwise, we would have a mere sequence of subjective representations *in us*, rather than an order *in the object*. The principle of causality is implied in the very notion of a succession inasmuch as it is posed as objective, and inasmuch as objectivity is defined by legality: "everything that happens, begins to presuppose something upon which it follows according to a rule."[10*]

We shall ask what is the nature of the object that one has in mind. A rule *qua* universal can define only an ideal object which, in principle, corresponds to an infinity of possible realizations. When we say that in a free fall the distance traveled is proportional to the square of the time that the object is falling, it is quite clear that there could not be a legitimate question of spaces and times that are, strictly speaking, 'given' in experience. The formula expresses a relation between two variables which are defined, obviously, on the conceptual plane. The particular event, as such, can never be part of an intelligible proposition whose being consists in its ideality. The author of the *Critique*, however, unquestionably understood reality in the proper sense of the term: 'what occurs', is the event itself that is given in an actual perception.

The reply will be that Galileo's law encompasses all possible cases: the mathematical formula allows one to calculate the totality of variations that can be encountered in reality; whereas the Aristotelian concept held to an abstract generality that was incapable of taking into account the diversity of things. But in our opinion, it is precisely here that it is necessary to distinguish in a radical manner the *real* object given in an intuition from the *ideal* object constituted by science. When I see a ball rolling downward more and more rapidly on an inclined plane toward the lowest point, I have the lived experience of an *event* which will never be reproduced as such, as opposed to the metronome which scans the measure of time. But I jot down in the 'book of my experience' to so many units of time corresponds a distance of so many units of space. This *fact*, even though it is 'given' in experience, is indefinitely reproducible. I have so little doubt about it that I don't even strive to assure myself of it: if I am going to attempt to repeat this operation, I try to vary the lengths to be measured.

These *facts*, calculated or corroborated by the scientist, are not always unique *events*, but ideal objects which are defined conceptually and are reproducible, in principle, in an infinite number of events. It is not enough to say simply that it is impossible to notice all the details of an experiment and that science must be satisfied with abstracting certain aspects of reality. One can imagine a concept of infinite comprehension which would encompass the totality of verifiable determinations in a given concrete experiment: it would retain the fundamental characteristic of the concept – viz. that it is capable of being realized, in principle, in an infinite number of occurrences. Yet, it belongs to the essence of reality to be posited at every instant in its irreducible particularity. To be sure, on the strictly scientific plane, it does admit of a structure which reflects a collection of possible predications. And even in the original perception, it does not present itself simply as a pure *this thing here* but its individuality as *this thing here* is due precisely to a concrete configuration which allows it to be grasped as a relatively determined unity within the horizon of the *life-world* (*Lebenswelt*). It remains that all these forms are grasped only in the ever unique actuality of their emergence. Consequently, the meaning of reality should not be reduced to a collection of characteristics, however rich or unified one might wish to characterize it. If, in analyzing two *events*, we find the same intelligible content, we shall conclude that the same *fact* is realized twice in history; but such a statement implies precisely the distinction of *events* wherein the *fact* is encountered. The event is perceived in experience; the fact is thought by means of concepts. And even though the concept claims to define the very *being* of the event (i.e., what it *is*) the latter retains, nevertheless, its radical originality, which consists in its *existence* and can be accounted for only by referring to the lived act of perception. The particularity of the real object is defined solely by the *temporality* of consciousness wherein it is constituted.

We understand now the disastrous ambiguity of the classical statement of the causal principle. The 'phenomena' to which it supposedly applies can be understood in two entirely different ways: one which leads to a contradiction; the other, to a tautology. If it is a matter of strictly *real* events, then the principle is manifestly *contradictory*: for an event is never reproduced, and it becomes entirely impossible within these conditions to define a rule according to which the 'same' cause always produces the 'same' effect. Law is possible only among facts. But then the causal principle becomes *useless*. Seeing that the fact is defined by its conceptual signification, the laws wherein it intervenes, will be nothing other than the conceptual predicates which constitute it. If these laws remain constant, then we have a *tautology*, since it belongs to the essence of a concept eternally to preserve its predicates. When it is stated that phenomena are produced according to constant laws, it cannot be a matter of events: for the event is never reproduced, and the very notion of constancy implies the concept of reproduction. It would be no more absurd than stating the principle instead of facts, since the fact (as conceptual meaning) necessarily remains the same in each of its realizations: thus it is useless to speak of constancy.

The principle of causality, however, bore an authentic signification hidden within the ambiguity of ordinary discourse. One ought not to say: "everything that happens presupposes something upon which it follows according to a rule," but rather, that events can be thought by means of concepts, considered as 'facts'. But such a passage could not be justified by the proof of the second category of experience, where it was only a question of analyzing the very notion of objectivity, in order to disengage the necessary rule from it. The 'objectivity' in question was that of the 'fact' as ideal signification, identical for all. The event exists essentially for the *contemporaneous* individual who perceives it. Its *objectivity* does not consist, yet, in escaping *subjectivity* by the suppression of its particularity within the universality of the law: it consists in being constituted in the *actuality* of an existing consciousness: the very time-lag implied by spatial distancing reflects upon that actuality. The real question is to understand *how that irreducible particularity*, emerging within a lived situation which does not refer, as such, to any determination in the objective world of science, *can be converted into an ideal objectivity, conceptually defined*. Yet, it is this very same problem which is *outlined* and immediately conjured away within Kantianism. The whole technique of the *Deduction of the Categories*, which Kant only reproduces in the *Analytic of Principles*, consists in substituting surreptitiously the *fact* (an ideal object identical for all) for the *event*, which is in an essential way relative to the temporality of the consciousness which perceives it. The originality of the intuitive synthesis was suppressed totally, to the advantage of the conceptual synthesis, thanks to a simple play of definitions. The analysis of the notion of *fact* ended easily in the notion of rule; for only a rule can define an ideal object. The result is extended to the event in favor of the ambiguity of the notion of object. Thus, one could, by means of a purely *analytical* procedure, justify a system of synthetic judgments, and the logical necessity of analysis concluded paradoxically to an *a priori* synthesis.

The original idea of Kantianism, which consisted in dissolving the phantasmagoria of metaphysics by grounding knowledge in the conditions of possible experience, thus lost its authentic meaning. This project implied a direct description of the world of perception – the *life-world*, in Husserlian language.[11] Yet, the *Critique of Pure Reason* continually substitutes the analysis of judgment for that of perception, in order thereby to shift the results of the latter to the former. Thereby, the signification of the whole system became immediately unintelligible. Inasmuch as the object of perception is perceived as 'determined', according to Kant, in the sense of 'being determined in relation to the logical function of judgment', it is clear that the meaning of objectivity shall consist in the forms of predication, and the proper role of intuition will be reduced, in fact, simply to furnish a manifold to the understanding, in order to enable it to particularize its laws. The necessity for experience derives, therefore, from a weakness of the human mind, rather than from the very structure of knowledge. A completed science would contain within itself the whole manifold, and recourse to intuition would become useless: the hypothesis is, no doubt, unrealizable, but the problem bears upon the principle, rather than the fact of it.

The demand for a reference to sensibility originally implied deeper intentions. It was not simply a matter of furnishing knowledge with a manifold, but of giving it its *meaning* of knowledge – as *knowledge of things*. The radical fault of the pure concept does not lie in its poverty but rather in its unreality: for giving it the richest content possible will never make it attain to existence. The critique of the ontological argument would not have any meaning whatsoever, if the understanding had to search for a mere manifold in experience. For the being which possesses the totality of the manifold – *ens realissimum* – does not, for all that, transcend the ideality of the concept. The task of actual knowledge does not consist in *particularizing* the category, but in relating it to an *existent*.

That the *critical idea* could not find an adequate realization in the *Kantian system* is what appears, evidently, in the paradox that the *necessary* solidarity of sensibility and understanding in the movement of knowledge results, definitively, from a double *contingency* – i.e., the *blind* character of *our* sensibility and the purely *discursive* character of *our* understanding – determinations posited in such an arbitrary manner that nothing could militate against the hypothesis of an intuitive understanding. Under these conditions, the deduction of the categories remained on the purely formal plane of the notion of object and could not ground a *material* agreement between the understanding and the actual givens of the sensibility in a manner that would guarantee the *real* possibility of a system of *particular* laws of nature: that the construction had taken its point of departure from two abstract terms, whose harmony was affirmed as necessary, will always appear as perfectly inexplicable. Furthermore, there remained even a recourse to finality, which was to be achieved with the *Critique of Judgment*. But this problem was posed only because scientific endeavor had been severed artificially by means of philosophical analysis into two moments of one and the same lived dialectic. Their union does not seem, therefore, to be capable of realization, except in the form of a perfectly arbitrary restriction. In fact, the limitation of knowledge to the domain of experience is not due to a contingent disposition of our mind: it expresses the very essence of the category whose total *meaning* consists in the actual act which constitutes the scientific object as the ideal unity of necessary determinations. That this act necessarily be accomplished on the foundation of the sensible is what belongs to its very actuality. The category taken 'apart' on the plane of the pure concept is revealed immediately as *bereft of meaning*. When the author defines it as *concept of the synthesis*, it must not be understood as if the concept existed from all eternity in the mind but finds an occasion to manifest itself only in the *shock* of experience. As a matter of fact, the purity of the category is so only because of philosophical reflection, which necessarily presupposes the concrete act, and this is only effective as taken hold of in the consciousness of self. The restriction of the understanding to a merely empirical usage is thus not an arbitrary limitation, imposed by the necessity of its being applied to a *given*, which for the rest of us human beings is found to be a sensible given; it results simply from the very signification of the category as philosophical abstraction, performed on the act which constitutes the world – not a world in general but this world, i.e., the world of actual experience. Thus the concept of formal unity reflects the lived consciousness of self, where the self is revealed to itself in its *transcendental actuality*. Such an actuality, inasmuch as it is *determining* and not *determined*, escapes all forms of objectivity: it, nevertheless, is posited in the irreducibility of an *I am*. "The *I think*," according to the text of the *Deduction*, "expresses the act of determining my existence."[12*] In this act " . . . I am conscious of myself, not as I appear to myself, nor as I am in myself, but only that I am."[13*]

This pure lived experience of spiritual existence, in its constituting act, could not find an adequate conception in the Kantian system. Inasmuch as the sole mode of knowledge envisaged in the *Critique* was knowledge of the physical object, the original self revealed in the consciousness of the self could be defined only as *unknowable*. "The consciousness of self is thus very far from being a knowledge of the self . . . . I exist as an intelligence which is conscious solely of its power of combination . . . ."[14*]

Nevertheless, this *unknowable* was not altogether *unknown*, considering that each person is constantly and most profoundly consciousnes of himself and, besides, that this consciousness is presented as the keystone of the *Critique*, taken in its entirety. Thus, this lived experience will be defined in such a contradictory manner that all knowledge of it becomes impossible. All knowledge implying the union of the concept and the intuition, the 'I think', will be *thought without intuition*, or *intuition without concept*. "This *representation* is a *thought*, not an intuition," and (the author adds in the *Analytic of Principles*), "the consciousness of myself in representation: the 'I' is not an intuition at all but a simple intellectual representation of the spontaneity of the thinking subject." But everyone knows the famous note of the second edition of the *Paralogism*: "The 'I think' is . . . an empirical proposition and contains within itself the proposition 'I exist . . . '. It expresses an indeterminate empirical intuition."[15*] This text appears perfectly clear to us. It can be concerned only with *transcendental experience*, in which the *determining* self is given to itself without finding itself thereby *determined*: seeing that every *determination* has been placed by the *Analytic* under the jurisdiction of the categories (which are, in fact, the categories of the physical object), the constituting act which escapes them can be grasped solely in an *indeterminate* intuition. Thus the concrete transcendental consciousness, which was to be thematized systematically by phenomenology, could be defined for Kant solely on the irrational plane of a *pure existence*. "An indeterminate perception (as the note which we are about to cite, specifies), here signifies only something real which is given, given indeed to thought in general, and so not as appearance nor as thing in itself (*noumenon*) but as something which actually exists, and which in the proposition 'I think' is denoted as such."[16*]

Thought without intuition or intuition without concept, the 'I think', designates the actual consciousness of self which inspires the general movement of the *Critique* without being able to be inserted in the architectonic of Kantian concepts: its purity could be preserved only by a set of prohibitions. Such a method is easily understood in the light of Kant's polemical position. Against the false objectification of traditional metaphysics, it was important to repudiate, in a radical manner, the possibility of transforming the *consciousness of self* into a *knowledge of self*, since every attempt at *knowing* the constituting act as such involved the risk of reifying it. The principal doctrine of the ideality of time was the main theme that would justify this negation. Seeing that all knowledge presupposes an intuition, and seeing that the intuition of consciousness is presented in itself in a temporal form, the ideality of this form would allow one to view lived experience as a simple phenomenon, justifiable in terms of *psychology*.

But it is enough to mention that this time has previously been entirely spatialized and that it is precisely by reason of these properties that were borrowed from space that its reality is denied, as a well known text of the Second Deduction explains:

> ... by the fact that we cannot obtain for ourselves a representation of time which, in addition, is not an object of external intuition, except under the image of a line such as we draw; and, seeing that without this mode of depiction we would never be able to recognize the unity of its dimension; and, similarly, seeing that we always derive the determination of length of time or of points in time, for all interior perceptions, from the exterior things we represent of the changing; and, consequently, seeing that we must arrange in time the determinations of inner sense (inasmuch as they are phenomena)

exactly in the same manner that we arrange in space those of external sense; and, consequently, seeing that we admit of the latter that they are known to us as objects (*Objecte*) only if we are also affected from outside – we, accordingly must avow that by means of inner sense we intuit ourselves only as we are affected inwardly *by ourselves*; i.e., in that which concerns internal intuition we know our own subject only as phenomenon and not as it is in itself.[17]*

Thus, the determining self has been posited 'outside of time' for the sole reason that time has been conceived as *constituted* according to the categories of the external object. Such a result does not exclude but implies the notion of a *constituting temporality*.

We can readily be convinced of this by examining the doctrine of the Antinomies. If we remain on the surface only, we find that the *Antithetic* of pure reason finds its solution in the distinction of objects as phenomena and noumena – a distinction that is itself grounded in the theory of the ideality of time. The dialectical opposition results from treating phenomena as noumena, which leads one to think of constituting syntheses as either finite or infinite totalities; but critical reflection discovers that these syntheses, inasmuch as they bear upon phenomena alone, must be taken only in their actual accomplishment as always unachieved: this unachieved character defines them in their meaning of truth – inasmuch as, while they always remain within the finite, they constantly require an infinite going beyond. But such a solution can only make sense if the constituting act itself be temporal. It is not just a matter of an appearance due to its insertion within the form of the inner sense. For if the act itself were atemporal, it could be thought to be achieved, even though it be given to itself as an always unachieved form; but such a concept is clearly seen to be contradictory. Hence, the unachieved quality does not simply characterize the psychological consciousness given in the inner sense, but, as well, transcendental consciousness itself in its constituting function: the world is neither finite nor infinite, because every assertion as to its size acquires a meaning only in reference to the actual accomplishment of the synthesis of its parts – which always remains unachieved. The solution of the antinomies rests, then, certainly, on reality – not, if you will, on 'time', but on temporality, since the unfinished character of the act constituting the world makes the discharge of the adversaries on equal terms fully evident. To say that one must not consider phenomena as things-in-themselves simply means that the work of constitution must not be taken in the abstraction of its *a priori* structure but in the actuality of its accomplishment.

Such, moreover, was the inspiration of the *Critique* in its fundamental theme. It was a matter of spurning the pretensions of dogmatism by showing the *regulative*, and not the constitutive, character of the Ideas of Reason. The concepts of classical metaphysics are meaningful only as rules of the understanding in the constitution of experience: to make them real was to steal them away from the very task which alone gave them a *raison d'être*. The whole critique of dogmatic rationalism aimed only at bringing to light the necessity of a concrete accomplishment that remained always unachieved, and was always to be pursued further. The limitation of the use of the categories to the data of intuition is, thus, less a restriction that one may judge arbitrary than a manner of defining the authenticity of the constituting act, inasmuch as this act itself exists only in its lived accomplishment. The *a priori* forms of the sensibility are but one aspect of this accomplishment, and the transcendental idealism deduced from it enables one to eschew the temptation of looking at truth as a ready-made thing, in order to engage in the real task of constitution.

The interpretation that we have just given of Kant's speculative philosophy may appear difficult to sustain for his practical philosophy. For the issue here is no longer to reason within the limits of experience but precisely to make reason go beyond these limits by forbidding it any sensible motive power; the value of reason *qua* practical lies in its *pure* usage. The analysis of moral consciousness, according to the *Foundations of the Metaphysics of Morals*, ends in "an intelligible world (i.e., a total system of reasonable beings as things-in-themselves)." The moral act, as free act, is performed within the atemporal, even though we have the intuition of it as a temporal phenomenon. Its rationality is not given to us in itself, but only in an indirect manner, in the form of the maxim of our conduct inasmuch as it can be erected into a universal law.

If we push the logic of concepts to the extreme, we shall say that the total reality of the moral life consists in an act accomplished from all eternity in an unknowable beyond, and that our existence in this world merely unfolds in a temporal manner. The very absurdity of a conclusion of this sort, which would render all moral effort meaningless, suffices to dispel such a literal interpretation. Kant's practical philosophy does not consist in referring morality to the inaccessible *in-itself* of a noumenal world, but rather in reminding everyone down here of the reality of its existence, seeing that by its very actuality it admits of a meaning of truth. The practical law is not an abstract form which has to be imposed from outside as the maxim of the will, but the expression of the moral intention in its lived intimacy: a good act, morally speaking, is not an act done *in conformity with duty* but an act done *out of duty*. But such an intention implies the actuality of its lived experience, and Kant's refusal to place liberty within the phenomena of the internal senses aimed precisely at preserving the purity of the consciousness of self in an act which exists solely in its actualization. The concept of intention escapes psychology, not because it would refer to an unknowable beyond, but because no psychological entity can be substituted for the moral effort which knows itself only by actualizing itself.

In this way, formalism, owing to its profundity, refers to a consciousness in action. The theory of the intelligible character, interpreted literally, would entail the negation of all moral progress. But this progress is required precisely by the Kantian concept of morality, and it remains only to make its authentic meaning more precise – i.e., that the moral effort does not tend simply to a greater conformity of conduct to the law, but rather to a radical overthrow of the very foundation of the maxims of the will. That the *becoming* of morality bears on its *intelligible character* as such, is the decisive proof that the doctrine of the noumenal world did not intend any immutable eternity but simply the profundity of the consciousness of self.

> But if a man is to become not merely *legally*, but *morally*, a good man . . . that is, a man endowed with virtue in its intelligible character (*virtus noumenon*) . . . *this* cannot be brought about through gradual *reformation so long as the basis of* the maxims remains impure, but must be effected through a *revolution* in the man's disposition. . . . He can become a new man only by a kind of rebirth, as it were, a new creation . . . and a change of heart.[18]

A philosophy can be viewed in terms either of its concepts or its intuition. Kant's concepts designate the conditions for the possibility of the object, such that these conditions are abstracted from the analysis of judgments which determine the object on three planes – theoretical, practical, and aesthetic. The transcendental philosophy thus seems to present itself as the system of formal structures which define objectivity in its different domains. But analysis had an actual value only because it was elaborated on the foundation of the lived consciousness of self in the concrete movement of the act which constitutes the world. Reference to this movement alone could eliminate the abusive reification of rational concepts in dogmatic metaphysics. Hence, we can say that the intention of the critical philosophy was to strive for this *transcendental experience*, in which the constituting consciousness is given to itself within phenomenological evidence. But this inspiration is translated within the system only in a negative and contradictory form in which the self loses that actuality which gave it its whole meaning. The phenomenological reduction, by thematizing the transcendental domain in the form of a concrete subjectivity, does not make us fall back into psychologism: it realizes in an authentic manner the original signification of Kantian idealism.

CHAPTER THREE

THE PROBLEMS OF REASON

13. SELF-EVIDENCE (*ÉVIDENCE*) AND TRUTH

The explications to which we have devoted ourselves to this point remained on a plane that was still formal. They enabled us to rid ourselves of constructivist temptations: but we still have to subject the transcendental domain to positive analyses. It is only here that the phenomenological task really begins with all of its difficulties. It is not simply a matter of finding the concepts which will allow us to dominate the *Heraclitean flux* of lived subjectivity; the constituting value of the concrete Ego, inasmuch as it ultimately grounds the objectivity of the object, must still be demonstrated factually. In other words, the phenomenological description must, owing to its very fidelity to the experienced given, elaborate the authentic solution to the problems of *reason*. We must not, thereby, understand just the traditional problems arranged under the title of the 'theory of knowledge'. The phenomenology of reason extends to all domains of being: it intends the whole meaning of experienced life – every right of consciousness to posit itself as true, to *live rationally and truthfully*. Every predicative or antepredicative experience, every intuition of values or ends implies claims that must be criticized and legitimated within the limits of their actual signification. A critique of this sort could not proceed according to abstract categories: it is realized by the very exactitude of the analysis, in which the presence of consciousness to itself guarantees its truth meaning. In a word, what is at issue is a *general critique of evidence* as *the lived experience of truth*.

The project of phenomenology described in this way, must now be justified in its content. How can truth be grounded in [self-]evidence when this very notion seems to exclude all possibility of mediation and thus of a rational demonstration? The reality of the object would be reduced to an irreducible *given*, which would have no other guarantee than that of an inner certainty of which each person is the sole judge for himself.

Husserl's reply is perfectly clear. All the difficulties come precisely from the fact that one has taken [self-]evidence to be some kind of internal sentiment, which, in a mysterious way, would serve us notice whenever some of our representations agree with the object: it is clear that such a criterion would be unthinkable, since nothing in the concept of such a sentiment would prevent its form being accompanied by false ideas – from which experience (in addition) would be sufficient to protect us. But *[self]-evidence is not a sentiment*. It "is not a psychic indicator affixed to a judgment and calling to us like a mystical voice from a better world: Here is the Truth! As though such a voice had anything to say to free spirits like ourselves and had not to make good its title to authority."[1] In fact [self-]evidence is defined not in terms of its 'real' content but of its intentional signification; it is the *primordial mode of intentionality*, the privileged moment in which *the thing itself* is presented – i.e., that very same one of which we speak and that we can no longer doubt, because every discourse constantly refers us to such a presence.

Suppose that I ask myself whether the person whom I saw yesterday has light or dark eyes: there would be no other possible solution than to try to see that very person, or to ask people who had retained a precise remembrance, or for me myself to make an effort to recall. In the last two cases I refer myself to a prior perception to the extent that it is *still present* in me or others. The whole meaning of the question is related to [self-]knowledge of this type; for, doubts are born from the fact that we are distant from the object in question, and all the problems that can be raised about it come down to asking how it would be perceived if it were there *in flesh and blood*. [Self-]evidence, more precisely perceptive [self-]evidence, is the primordial mode on which all other modes of intentionality necessarily depend, inasmuch as what they intend (in a more or less obscure or empty fashion) is none other than *that very thing* which is given fully in itself.

We may note that our sketch just given for the elementary example of sense perception has, in fact, an absolutely universal import. Thus, a theorem of geometry can be the object only of a *confused intuition* or of a *blind or symbolic representation*. The first case occurs when I have an obscure remembrance of the demonstration, or when, being ignorant of it, I divine its general structure in a more or less confused feeling. The second case is that of a purely verbal comprehension, in which I have no idea of what steps to follow. Doubt is then permitted. But the only sensible thing to do is to subject it to a careful analysis, in which the theorem is presented in the fullness of *itself* by the performance of the operations which demonstrate its truth. The notion of intuition does not exclude that of mediation: the particularity of predicative contents can rightly be given to intuition only through the mediation of the operative syntheses of judgment and reasoning. When these syntheses are performed in a neat and precise manner upon the actuality of lived experience, we possess the *primordial [self-] evidence* – the *perception* of the theorem. [Self-] evidence is *non-primordial* when the demonstration is *still present* in a clear and distinct remembrance, even though not retained in detail, or when we content ourselves with following an exposition which we *understand* perfectly but which we do not actually repeat in itself. It is clear that *mathematical truth* does not signify anything more than precisely what is given in [self-] evidence of this kind: in it is found the solution to all conceivable problems.

We can perform the same analysis within the domains of feeling and action. The truth of love is defined by an 'experience' in which the object is revealed *as evidently and obviously lovable*. [Self-] evidence is *primordial* or *perceptive*, when it arises in the freshness of first discovery, or when it is fully renewed in the presence of the loved person. There is still self-evidence, though nonprimordial, when love lives off the *ever-present* memory of the impression which gave birth to it, or when the feeling is motivated only indirectly, as when we love a person *for the love of another person*. Doubt appears only when the feeling is confused, or when distance or habit have weakened the remembrance of the original *perception*: in all these cases there is still an *intuition*, but it is *not [self-] evident*. Finally, love will be purely *verbal* or *symbolic* if it is grounded in duty or self-interest, or even in imitation or as a worldly game: these are, indeed, the same values which are intended here, but consciousness does not grasp their actual content – albeit in the form of a confused and obscure intuition. Naturally, to the extent that we get further away from primordial evidence, the chances of error become multiplied. Love can be shown to be false in an experience in which the object which we believed we loved, is exposed as obviously hateful. Finally, within the domain of action, truth is defined as practical [self-] evidence, in which a vocation or duty is revealed as that from which we cannot escape. Here again, the reference to lived experience does not imply any 'psychological subjectivism'. An end has an actual value to the extent that it is *perceived* in the experience of *praxis*: disputes arise precisely where either the action is guided by confused memories or the signification of a purely verbal comprehension of rules has not been tested.

Such, then, are the fundamental modes of intentionality: primordial [self-] evidence ('perception' or 'experience', in the broad sense) and non-primordial [self-] evidence, derived by means of the clear memory or comprehension of the other – a confused intuition, an empty or symbolic representation. These modes, which imply the status of the object as existent, are doubled in corresponding imaginary modes. Instead of performing the demonstration which enables me to obtain the *perception* of a theorem, I can content myself with *imagining* it, by representing to myself the process which I could follow in terms of what I know of the general structure of the mathematical domain. In fact, every speculative discovery is imagined before being actually verified (*'perceived'*). Such a consciousness is based on clear memories some of which in their turn can simply be imagined: thus I can suppose that I have already performed certain demonstrations and, on the basis of these imaginary remembrances, begin to outline new structures – always within the imagination. In the same way, I can train myself to imagine a confused intuition or a symbolic representation. We can easily find corresponding examples within the domain of values and ends. In the preceding analyses, we ourselves have not really lived the intentionalities which we have described: we have contented ourselves with *imagining* them.

But it is quite clear that the imagination derives all its meaning from its *own* thetic acts of consciousness. These thetic acts of consciousness, in turn, by positing the existence of the object, necessarily refer to [self-]evidence, in which the object is grasped in its very presence. Acts of consciousness could not give it a particular definition but precisely intend it *just as it would be revealed, were it to show itself.*

Thus, all the ways which consciousness has of relating itself to the object, unanimously refer to [self-]evidence as the *primordial modes* of which they are but *intentional modifications*: it is this very reference which constitutes them in the meaning of their being. "*Intentional* modifications have quite universally the *intrinsic property* of *pointing back* to something unmodified."[2] [Self-]evidence is thus not simply an unjustified and unjustifiable 'given': it realizes in itself the precise meaning of every possible justification. It is justified by the very movement of constituting life, inasmuch as it consists in returning repeatedly from the modified to the primordial. Disputes and problems which arise when one is satisfied with *speaking about the thing* by representing it in a more or less distant manner, have no other signification than to refer to [self-]evidence where what was obscurely felt or 'simply intended' is presented in the plenitude of *itself*. In a word, it is error itself which bears witness to the truth, since the obscurity which engenders it implies in its constitutive meaning the procedure that ought to dissolve it. The whole being of the confused consciousness consists in an aspiration to [self-]evidence, where the aspiration is suppressed by realizing itself: *falsum index sui et veri* [the false is an indicator of itself and of the truth].

14. THE PROBLEM OF ERROR

It was relatively easy to legitimate the notion of [self-] evidence as the primordial mode of intentionality, by opposing it to unauthentic acts of consciousness. A more delicate problem is raised if we examine the cases in which error does not (in a manner of speaking) denounce itself by its intrinsic confusion but brings itself forth out of its [self-] evidence. It is precisely here that the phenomenological method reveals its efficacity: since [self-] evidence is not a passive given but a privileged moment in the becoming of constituting life, it could not definitively guarantee the truths which it discloses. On the contrary, it belongs to its essence to be able to invalidate and correct itself. Self-evidence could never be obtained if it had to posit the absolute of a *truth in itself*. But it is precisely the merit of phenomenology to have discarded every prejudice of this sort.[1]

When I perceive an object, I can assert with total certitude the properties which I grasp of it: their reality consists, precisely, in being given in this manner. But it is revealed in the course of experience that my perception was false: I believed that I had seen a man when in fact it was only a dummy. Such a discovery does not diminish, in any way, the [self-] evidence of perception as such: for it is precisely on the basis of an actual perception that I invalidate and correct my previous perceptions. That a force be overcome by a greater force does not imply any contradiction but follows from the very notion of force.

The observation just made has an absolutely general import. The possibility of error is not confined to the case of sense perception. The claim of [self-] evidence is but the claim of actuality as such: it passes at each instant to new evidence – each [evidence] retaining this privilege only within the confines of its actual lived experience. Hence, it belongs to its very *meaning* to be able, once having elapsed, to be revealed as false. This is valid for all [self-] evidence or every 'experience' (in the broad sense). "Even a self-evidence which is offered as apodictic can be unmasked as an illusion, which presupposes, nevertheless, an evidence of the same kind to 'shatter' the former self-evidence."[2]

Thus, since philosophical propositions always are postulated on the level of necessity, we can say that philosophy, to the extent that it is *true*, rests on apodictic evidence. This in no way prevents the self-contradictoriness and self-correctiveness of diverse doctrines nor besides, their philosophical value, from being contestable. The very evolution of Husserl's thought provides us with a good example of this. The *Logische Untersuchungen* had demonstrated, with every desirable clarity, that truth possesses an ideal being, which is radically independent of the lived experiences by which we accept it: it belongs to its very *essence*, that without which it would be totally inconceivable. Yet this proposition, which can be verified by means of apodictic evidence, will be revealed as false, at least in its exclusivity (and rightly so), when transcendental analysis will have shown by means of a superior apodicticity that it belongs precisely to the essence of all truth to be constituted in the pure becoming of concrete subjectivity. Truth is no longer an *in itself* but the product of the actual act which engenders it in the absolute flux of transcendental life. Nevertheless, it is still the case that the ideal formations thus constituted are posited as eternal and universal – with the consequence that the first proposition is realized in its authentic meaning only by being suppressed in the second. Transcendental experience itself, which appears after the reduction as the supreme form of apodicticity, is, nonetheless, susceptible to being revised and corrected at every instant. New aspects are discovered, which subvert the meaning of former evidence. Thus, the *genetic* structure of the Ego, such as it appears in the writings of the last period,[3] contradicts Husserl's initial position in the *Ideen*, as the identical pole of lived experiences of pure immanence. But these successive reversals within the doctrine do not diminish the apodictic value of the *phenomenological experience*: on the contrary, they imply its absolute self-evidence in the actuality of its achievement.

Classical thought, by interpreting truth as an immutable *in itself*, was excluded from understanding the movement of actual knowledge. For, if the object be an absolute, there could not be a middle-term between knowledge and illusion. Either consciousness exits only in its actual lived experience, which means that its errors of fact denounce it as purely psychological, and knowledge will become an impossibility. Or it is in the nature of thought to be thought *of being*, and then knowledge is fully justified, but error becomes inexplicable, since it would consist in positing nothingness as being, yet since it is nothing, it should not even be conceivable. Knowledge is absolute, or it is not – such is the dilemma which was imposed within the horizon of classical concepts, in defiance of the experience of actual knowledge, which shows it to be a perpetual becoming, whose truth is not suppressed but realized in the very relativity of its moments.

Difficulty arose from the dawn of philosophical speculation, when the very constitution of the idea of truth as absolute truth revealed error as unthinkable; and we know how Plato, after having made being the unique object of philosophical contemplation, was led to affirm the existence of non-being, in which would consist the precise being of error. A formula of this kind readily allowed for the enunciation of the problem, but not for its resolution. For, non-being was accepted solely *by virtue of being non-being*: it *is* what the object *is not* and thus can be attributed to the object only *as what this object is not*. But, then, one does not see how it could be identified with being – i.e., that which rightly defines the fact of error: to be mistaken, is not to attribute to the object its non-being *qua* non-being, but rather *qua* being. If Theatetus is sitting, the act of flying will be part of what he is not – in other words, of his non-being. But since non-being has been justified solely *qua* non-being, the theory, in fact, explains only that we can say *Theatetus does not fly*; it in no way helps us to comprehend how this non-being could be affirmed as being in the proposition: *Theatetus flies*.

The existence of non-being *qua* non-being or as 'other than being' does not raise any difficulty. What matters, however, is to give an account for the appearance which presents non-being in the form of being. The Platonic analysis, remaining on the plane of the object, did not throw any light on this problem. For, all one could say of the object was that it is what it is, and is not, what it is not: yet, the Sophist claims that it is what it is not. Error manifestly reflects upon the intentions of consciousness, and this is precisely the conclusion that Plato reached in his famous metaphor of the pigeon-house. But it was hardly possible to express this conception within the objectivist framework of the philosophy of ideas. Modern thought, since Descartes, resolves the problem by acknowledging the strictly subjective plane, with the notion of a pure will which can choose in the absence of a clear and distinct knowledge. If I hold to my actual perceptions, I will find that I know some things while being ignorant of others, but nothing enables me to posit my ignorance as knowledge. Thus, the act of positing is independent of perception itself, and necessarily refers to a *freedom*. Hence error is a fault, and it is sufficient that we make a good use of our free will and make sure not to go beyond the actually perceived given within judgment in order to be assured of never being desceived.

But if freedom is given a rule in this manner, it itself can no longer be defined in terms of the purely arbitrary. Descartes admits specifically:

in order for me to be free it is not necessary that I be indifferent as to the choice of one or the other of two contraries; but, on the contrary, the more I lean to one – whether I recognize clearly that the reasons of the good and true are to be found in it, or whether God so disposes my inward thought – the more freely I choose and embrace it . . . . For, if I always recognized clearly what was true and good, I would never have trouble in deliberating as to what judgment and choice I should make, and then I would be entirely free without ever being indifferent.[4]

There is thus at least one case – that of clear and distinct knowledge – in which the will could not be separated from the understanding: for freedom is realized by the very necessity of [self-] evidence. Hence, it is no longer possible to give an account of assent as a free choice. If we recall that confused perceptions themselves imply a positive content – for to perceive nonthingness is not to perceive – and that, on the other hand, they always give way before clear and distinct knowledge, we will be obliged to conclude that it is the objective content which, in all cases, determines adherence, since it is implied in proportion to the richness of its content. If the assertion of the will were really independent of the perception of the understanding, distinct knowledge could never be imposed; but the fact is that it is imposed – thus proving that the act of assent follows from perception itself. But then error could no longer be explained in terms of freedom; it simply results from insufficiency of knowledge, inasmuch as it deals only with part of its object. In this way we return to the plane of the object, and the historicity of the knowing subject is once again reabsorbed within the system of science. Since all thought is thought *of being*, error will have nothing positive in itself: it will be but a moment of truth – and necessarily true inasmuch as it is a moment.

The end result of classical thought, according to Spinoza and Hegel, verifies the presentation that we gave at the beginning of our last paragraph. To posit truth as an absolute is to condemn ourselves to require of knowledge that it, at one and the same time, be absolute, under pain of not being knowledge: thus, error as such becomes inexplicable. Undoubtedly, some positive content can be given back to it by referring to the pure act of subjectivity. But seeing that the conception of truth as an *in itself* was placed entirely on the side of the object, there no longer remained the abstract concept of liberty of indifference to define subjectivity as such, and the requirement of the content entailed an immediate return to the object: error is dissolved in knowledge, which from then on justifies it as one of its moments.

Yet, it is the case that the positing of absolute knowledge is self-contradictory. For, if one defines truth as the totality of a process, each moment of which will be abstract, and if one takes that moment in itself (false in this abstraction but true within the whole, since it belongs to be very *meaning* of the process never to be completed), we never shall have anything in fact but abstractions posited as totalities; in other words, we shall always remain, in fact, in error. Dogmatism necessarily ends in scepticism, each based on the definition of truth as obscure.

But it is precisely the intrinsic absurdity of this common postulate which informs against phenomenology, realizing thereby the profound intentions of the Kantian critique by ridding it of its formalist prejudices. Truth, being defined in terms of [self-]evidence, cannot by that very fact be posited in the absoluteness of an *in-itself*, for it belongs to the precise *meaning* of all [self-]evidence to be fallible. [Self-]evidence exists as such only in its actual lived experience, which is renewed at each moment. Such a movement implies an absolute privilege of present actuality over the results previously acquired and the right of constantly rectifying them. Thus, truth is defined only in terms of its *becoming*, which must not be understood as an intelligible movement of ideas, but as an actually lived temporality. Constitution-consciousness is a *Living Present*, which constantly unfolds, and each moment of it in its turn acquires the privilege of actuality. But none of them as such can be posited as absolute, since the claim of each belongs only to the instant of its realization. The present continuously passes into the past and keeps its meaning and value to the extent that it is in accord with the *new present*. Thus, the possibility of error is implied in the very movement of consciousness in which truth is constituted. The meaning of evidence, as grasped of the *thing itself*, implies the absolute right of the present actuality to correct results acquired in past evidence. Error is not an obstacle or a limit to knowledge: it results from the very carrying out of its truth-meaning.

These remarks do not, in any way, constitute a 'theory' of evidence. They make explicit only the meaning of notions. Each of us understands by truth what is placed before us *at present*. But the present remains present only by always becoming another. Thus, truth could not be defined by the absoluteness of an immutable *in itself*: but it is only absolute to the precise extent that it is constituted in the eternal becoming of the *Living Present* [*Lebendige Gegenwart*].[5]

15. [SELF-]EVIDENCE AS INTENTIONAL PERFORMANCE (*INTENTIONALE LEISTUNG*)

To this point we have considered [self-]evidence only in its formal notion, as grasped from *the thing itself*. We need to make this concept more and more specific, in order to undertake the task of positive description. [Self-] evidence, as the primordial mode of intentionality, is defined in terms of the particular form of its phenomenological structure.

Seeing that every consciousness is consciousness *of* something, we must distinguish, first of all, the moment when it posits the object and 'characterizes' its existence in such and such a manner. Thus, perception grasps its object as *present*, memory as *past*, expectation as *future*. In contrast to perception as *presentation*, memory and expectation are *presentifications*, since they can render their object present only with the characteristics of the past or the future. This presentification can itself be *neutralized* and related to a time which is not ours and whose reality we refrain from affirming: we then have the imagination which characterizes its object as *imaginary*. In another direction, we find various modalities of the status of existence: immediate certainty is modalized in doubt, supposition, presumption, interrogation, negation. Correlatively, the object will be 'characterized' as existent, doubtful, possible, probable, discussable, nonexistent. Finally, the intentionalities proper to feeling and action give the object the 'characteristics' of value and purpose.[1]

These 'characterizations' presuppose, clearly, the representation of an object. The intentional act implies a moment by which it is related simply to the object as such: it is intentionality or the intention, properly so called, inasmuch as it defines the *noematic meaning* or the meaning of the intended object *qua* intended. It is the object and its determinations such as they are intended by me, which is found posited with the 'characterizations' just mentioned.

I see a tree. It presents itself to me with a certain number of determinations which define the objective meaning that it has for me (i.e., *what it is for me*). The object perceived in this manner is characterized as present. It would possess the character of past in memory and of future in expectation. If I see it in a fog, it will be posited as the same object, but characterized as doubtful or probable. Love and desire can give it the character of lovable and desirable. We note that these 'characterizations' are not objective determinations. The object is not lovable in itself, but the act of love colors it, so to speak, in such a manner that it is *experienced* and *felt* as lovable, without being objectively intended [or planned] as such. But the 'characterization' can change into the noematic meaning, because of an act of reflection which objectifies it: the object is enriched with a new determination and is thus made available for a new characterization. The loved one, *recognized* as objectively lovable, can give birth to a renewed love, in which it no longer is simply loved, but loved precisely with the very qualities that love has found in it. It also can, in certain cases, provoke jealousy, which will *characterize* it as hateful. Love is preserved, but only by virtue of being a feeling already acquired and immediately repressed. It no longer springs directly from the representation of the object, but is presupposed and absorbed in this very representation, inasmuch as it determines its object as lovable, from now on: it is upon this *determination* that the *characterization* of being hateful is constituted in a feeling of jealousy that is experienced spontaneously.[2] By the same token, the doubtful as such can be intended objectively in a reflection which verifies the reality of the doubt: but the object thus determined will be characterized no longer as doubtful but, indeed, as certain: it is *certainly doubtful*. The past is lived as past in memory. But intended with the objective determination of being past, it will no longer be characterized as past but as present: it is *presently past*.

The indications that we have just given, enable us to outline the general structure of intentionality in its two essential moments: *determining* and *characterizing*. This structure is itself constituted on the foundation of a *given* which, in its ultimate stratum, does not imply any intentional meaning. The object is posited according to the manner in which it is given: represented with an identical content of determinations, it can be grasped in a more or less clear or obscure intuition, even, likewise, in a perfectly empty one. I see this tree as a unity of objective determinations. But I actually grasp only certain aspects of it; more precisely, I have within myself a collection of sensorial data which, animated by the meaning of my intention, represent a tree for me. But this *representational content* (*repräsentierender Inhalt*),[3] which gives an account of the status of the object, can become blurred to the point of allowing for a mere presumption or supposition: thus, when I see the tree from afar or in a fog, I may doubt its existence. Consciousness intends the object according to a signification which is not always *fulfilled*: it can even be presented in an empty form as when the *representational content* is reduced to a verbal symbol. If I content myself in uttering the word *tree*, without perceiving, remembering, or imagining any object of this kind, I still have a representation whose meaning can be the same as that of an actual intuition, but an empty representation, since the 'given' is reduced to a *flatus vocis*, which represents nothing of the object *itself*. The variations in the *mode of fulfillment* account for the manner in which consciousness will characterize the status of existence: as certain, doubtful, possible, etc. . . . .

To be sure, in this outline we have given only titles of chapters, rather than actual analyses. They, nevertheless, enable us to understand already, in a more precise manner, the classification of modes of intentionality – i.e., primordial or percpetual evidence, evidence derived from the clear remembrance of the comprehension of the other, confused intuition, blind or symbolic representation. Henceforth, these distinctions refer to the moment of *fulfillment*. The rationality of the act of positing is grounded in the manner in which the *intended* meaning is actually *fulfilled* or, in other words, in the nature of the *'representational content'*.

For the sake of clarity of exposition, we have presented the fundamental phenomenological concepts in a schematic form, which can produce a type of uneasiness: have we not grounded the value of knowledge and, more generally, all truth-meaning in mere *data* and, in the final analysis, in *sensorial data*? Would not the doctrine of 'fulfillment' be reduced thereby to a new form of sensuous empiricism? The 'noematic meaning' would be a sort of framework which would be fulfilled by sensations or phantasms. Yet it is this very theory which must justify [self-] evidence as the ultimate source of all truth and rationality.

In fact, the objection comes simply from taking concepts in their abstractive capacity, separated from the *intuition* upon which they are constituted. The three moments which we have distinguished (i.e., a characterizing position, the intentionality of meaning, representational content and, correlatively from the noematic side, characterization, intended meaning, and given aspect) are not, so to speak, parts whose combination would form (on the former side) the experienced act and (on the latter side) the object posited as such. Their distinction does not refer to some sort of spatial strcture: it operates dialectically in the *temporality* of intentional life. The articulations of consciousness outline the projection of the transcendental becoming, in which the object is constituted in primordial evidence, which immediately becomes blurred in a more and more confused intuition, thereby falling into the emptiness of a simple symbolic representation – this latter, in its turn, initiating an inverse movement in which the intended meaning is rediscovered in the plenitude of new evidence, in which it can be enriched, confirmed, but also corrected or even suppressed.[4]

It is precisely from this movement that the moments which we have distinguished are detached. Primordial evidence is not a composition constituting a status of existence, an intentional meaning and a content which fulfills it: it is, in its totality, a presentation of the *thing itself* in its plenitude. But perception immediately passes to retention. Clarity diminishes as the object is *characterized* in a less and less distinct manner, while always being recognized within a synthesis of identification as *the same* object. A movement of this kind is possible only if, from the very beginning, meaning is postulated as independent of the intuitive given which supports it, so that it can remain identical to itself, outside of this given. And seeing that modes of appearance become more and more obscure and, consequently, terminate in an empty representation, the subject is tempted to realize this meaning in a thing which would be sufficient in itself. The intuitive givens then would come simply 'to fulfill' a ready-made form. But examination of the actual content enables us to dissolve such an illusion. A meaning is presented in an obscure intuition, or an empty representation can be defined in no other way than by reference to the evidence in which it would be presented in the plenitude of *itself*. The identity of full and empty meaning signifies nothing but a reference of this kind. If in a confused way I recall a landscape that I saw last year, I intend *the same* objective *meaning*, not in the sense that I actually and concretely would possess in the intentionality of recollection the same richness of determinate content which I had in perception, but simply in the sense that I would be conscious of intending one and the same object to which I can return – whether by returning to the same place or by interrogating my memory in a more precise manner or by having recourse to the testimony of another. But then, of course, the *actual* content of determinations is no longer the same, even though it is posited as being the same: the meaning, in fact, has been impoverished, and the return to clear and distinct intuition allows for precise completion of the recollection and, on occasion, its correction – even, to be sure, its suppression as mere illusion. Thus, the difference between full and empty meaning does not consist simply in the presence or absence of the intuitive given, but *the meaning is truly itself only to the extent that it is constituted on the basis of the given*. The authenticity of the movement is determined with regard to the status of existence. The moment of 'characterization' could not be effected apart: it defines only the manner in which the *passage* from the given to the intended is accomplished. Hence, the structure of intentionality in no way implies a composition of parts: it outlines the form of becoming of subjectivity in which the object is constituted in the meaning of its being.

We have seen, in a preliminary outline, that the distinction of intentional modes – of primordial evidence – from empty representation, referred to the moment of 'fulfillment'. Verbal appearances could have led us to the belief that it was a return to empiricism. We see now that the concept of 'fulfillment' does not presuppose a ready-made meaning that only needed to be 'fulfilled'. To be precise, it is a matter of *constituting* the meaning within the *plenitude of its being*. Thus, evidence could not possibly be defined as a passive given: it is the performance (*Leistung*)[5] of the constitutive meaning of the being of the world. The fact that this can be accomplished only upon the foundation of a *given*, whose nature varies according to regions but which always implies a sensorial kernel (if only by virtue of being a suppressed moment), is, accordingly, a decisive proof that the *world of truth*, engendered in self-evidence, is indeed *this very world* in which we live. Nevertheless, it is still the case that *truth* is produced only as *truth* by virtue of *transcending* the properly *'given'* elements, whose *true meaning* it rightly defines. Thus, the genesis of significations in experienced [self-] evidence, certainly deserves its name: *transcendental genesis.*[6] The consideration of temporality simply signifies that truth is not presented as a thing which is recognized by a sign but as a *movement* which takes its value only in its actualization. Only such a conception can justify the spiritual act, not in the abstraction of its theoretical ideal but in the actuality of its accomplishment. Hence, the shortcomings which we have demonstrated in the becoming of the spirit do not suppress the value of its claims: the possibility of error is implied in the very meaning of truth, as *truth of this world.*

16. THE POSSIBILITY OF ERROR AS CONTEMPORANEOUS WITH TRUTH

Self-evidence *produces* truth by positing the *meaning* of acts of consciousness which intend it. It is an available acquisition (*Erwerb*), which can be used by being satisfied with a more or less confused recollection or even a simple symbolic representation. It, likewise, will be possible to transfer it by a process of *association* to situations similar to the one which elicited the original evidence, and, in this way, to perceive the new object as an object of the same kind.[1] In this whole movement, the meaning is intended as remaining the same. But the actual content of determinations is modified by the acts which intend it: contradictions can be introduced in spite of the self which, in fact, posits something other than it thinks that it is positing. Thus, it happens that, after having understood a theorem perfectly, which is contained within a primordial [self-] evidence, we apply the theorem in a defective manner by taking it, on the suggestion of the concrete situation, with determinations that it does not have, *while attributing the same meaning to it, in theory*. The possibility of error is based on the opposition between the rigidity of meaning in its *avowed claims* and the malleability of its *actually real content*.

We note that a difference of this kind does not arise owing to a fault that one could avoid or to an inherent weakness of human nature. The meaning is constituted within the evidence as the *being* of the object – i.e., what it *is*, independent of acts of consciousness which intend it. Its rigidity is not the result of an aberration, but of its very position as *true* meaning. The act of knowledge is true only to the extent that it creates a valid signification that is independent of the subjective givens of the particular moment and whose true *meaning* it defines precisely. But, in virtue of that very creation, error becomes possible. The constituted meaning is preserved in acts of consciousness which imply a content that is different from the one presented in the primordial act. Indeed, its position as *truth*, from the very beginning gave it the right to continue to be affirmed beyond the evidence which engendered it. Thus, its very *meaning* allows it to be posited validly in an obscure intuition or a symbolic representation. But, precisely for this reason, many confusions and contradictions arise: the possibility of error is found to be implied in the very act that constitutes the real in its meaning of truth.

Thus it appertains to the meaning of every geometrical proposition to be posited as independent of the sensible intuitions in which it is represented. Such a meaning refers to a *primordial creation,*[2] which (inasmuch as it considers in the sensible given only what can be construed through concepts) projects an infinite horizon of possible operations in which the being of the geometrical is defined as the *truth* of real forms. Naturally, it is not a question of a framework that is given once and for all, but of a primordial signification which is *reconstituted* in each geometrician, inasmuch as all geometrical reasoning implies a negation of the sensible figure as such and the figure's absorption in a rational construction. That a dialectic of this kind can always be repeated is postulated in the very meaning of every geometrical proposition to the extent that it is affirmed to be *true*. But this very possibility opens the door to error. For the movement thus *presupposed* no longer needs to be reenacted at each instant in the authenticity of the original procedure. The geometer is satisfied to be handling his concepts as self-sufficient symbols. It is at this moment that sensible configurations are reintroduced without his knowledge, when he believed that he was dealing only with geometrical constructions. Error finds its source in a *forgetfulness* of the original act which created the domain in which present activity continues to give it the meaning of its being. Such a forgetfulness is not produced by simple inadvertence: the possibility of it is implicit in its very meaning as *always valid* meaning.

Given the rectangle ABCD (Figure 1), I draw in B, outside the rectangle, and make with BC small angle α, the segment BE = BC. I join DE. I draw the mid-points M and M^1 of DC and DE, respectively. M and M^1 intersect at I. We can see immediately that the two triangles IAD and IBE are equal, since their three sides are equal. Thus, their angles in A and B should be equal. Yet, it is obvious that they are not, since the angle IAD is composed of an acute angle IAB and of a right angle, and since the angle IBE includes the same elements (IBA = IAB and ABC = one right angle) and in addition the angle CBE = α.

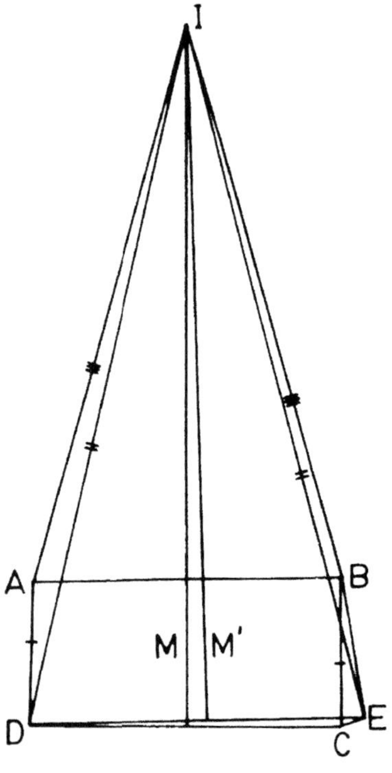

Fig. 1

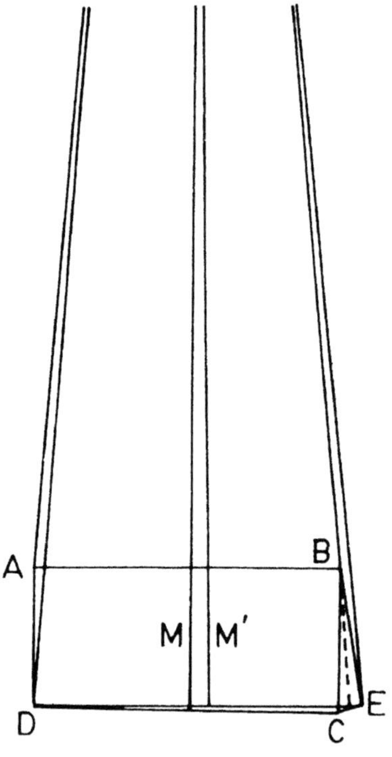

Fig. 2

The contradiction comes from the fact that to an intuitive configuration the value of a conceptual construction has been given. For since IB is on the mid-point of CE, the angle IBE is necessarily an exterior angle (Figure 2).

Yet, to draw according to the normal criteria of sensible intuition is to end up, regularly, with Figure 1. And the procedure may appear to be legitimate, since it is *understood* that no account will be taken of intuitive configurations as such, but only of conceptual determinations which are represented therein. But it is precisely this '*understanding*' which allowed for the construction to deviate from its conceptual rule. A slight inexactitude seemed unimportant because 'it went without saying' that the drawing was not intended in its sensible content. But this was taking rational determinations for realities *in themselves*; while their rational value rightly had its meaning only on the *foundation* of the sensible, by an *actual* movement which suppresses this sensible and elevates it to its *truth* in a construction realized exclusively by means of concepts. It is the dialectic of this passage which is both *preserved* and *forgotten* at the same time in error. It is preserved in a symbolic form and gives to the represented figure the meaning of a geometrical figure, and it is precisely the reason why we have not allowed a contradiction that would not have existed for naive perception. But while believing that we were doing geometry, we were satisfied in *presupposing* the procedure which thematized its domain, by *forgetting* to do it over again in a real manner; and the evidence proper to sensible intuition had concealed the rational evidence of the geometer. Error consists in a regression which can be defined only in relation to the superior level to which one has already elevated himself and which one continues to maintain. The possibility of it is obtained from the very moment of the first creation of the new domain, inasmuch as it is presented as a horizon of objects whose *reality-meaning* is possessed from then on – the task of knowledge no longer being to enter into the details of particular forms. The constitutive dialectic of the domain, as *passage* from the sensible to the intelligible, is thus forgotten as soon as it is achieved: its very accomplishment implies this forgetfulness.

We now see in what sense it is appropriate to understand the traditional theories that regard error as a moment of truth. Within the viewpoint of classical concepts, we were condemned to resolved the problem by the suppression of one of its terms. If error is a partial truth, it will be, in fact, truth and not error. For it is legitimate that each moment be affirmed as a totality on its own level: therein lies its truth. And if it is absorbed, thereafter, in a more elevated concept, it will always remain true, since it is posited therein as a transcended moment. The fact remains, however, that error does exist with a positive content that makes it error.

Actually, falsity is indeed a moment of truth, since it consists in the resurrection of a kind of evidence that has already been transcended but which, nevertheless, remains true on its own level. Thus, the numerous mathematicans who believed that they succeeded in demonstrating Euclid's fifth postulate were merely trusting in an intuitive evidence whose value is indubitable in ordinary life. But their error is not defined simply by this limitation: it implied the insertion of this sensible structure within the geometrical perspective and positing it as rational truth. The false is, accordingly, neither an isolated part in a system of eternal truths nor an abstract moment in an eternal dialectic of pure concepts. It arises in real temporality through an inauthentic act which retains from the original act only a claim that it does not succeed in realizing.

Thus, when we imagine that the sun's distance from us is about two hundred feet, our error does not consist as Spinoza believes, in that "while we thus imagine, we do not know the sun's true distance or the cause of the fancy."[3] For here, in the absence of any astronomical knowledge, we possess truth proper to the sensible world that can be verified by the agreement of men with normal vision. It is opposed on its own level to the error which would consist in seeing the sun ten feet away. The normal data of perception can thus be condemned as erroneous solely by virtue of a *pretension* to scientific truth – a pretension that we all affirm as the heritage of the original act that created the perspective of the world of science as the *truth* of the life-world.

It was, then, for good reason that Descartes disclosed an intervention of the self in error. But he was wrong in defining subjectivity in terms of abstract freedom, as a purely arbitrary choice. Hence, error became an inexplicable aberration, and Spinoza could object, rightly, that there is no adherence without motive and that the false can be posited only in terms of a particular meaning of truth. But it could not be the case of an objective truth that, no matter how partial or abstract one might wish, it would be in itself true and not false. The positive being of error can consist only in the meaning of the intentional act which, fully realized in the absolute evidence of the original creation is maintained in a *sedimented* form, as a confused or even completely empty pretention, but which, nevertheless, retains a truth value, inasmuch as it refers to the authenticity of the creative moment and thus implies the possibility of a *return to the origin*. In this absolute movement in which consciousness still remains present to itself by forgetting itself, and forgets itself only in order to rediscover itself, time is constituted as the other in which the moments of a *history* succeed each other. Existence in time with its limitations and failures expresses only the ever living presence of the original act where its *horizon of truth* is constituted: in this manner is explained the ambiguous nature of which Plato spoke, when he had to recognize the being of illusion.

17. A DIGRESSION – THE THEORY OF EVIDENCE ACCORDING TO DESCARTES AND THE PROBLEM OF THE CARTESIAN CIRCLE

If the *cogito* is presented as the immovable rock which must support the edifice of knowledge, it can do so only by being explicated within a theory of evidence. And, in fact, this first truth acquires its actual meaning only from the rule that Descartes abstracts from it in order to define all truth in general.

> Certainly in this first knowledge there is nothing that assures me of its truth, excepting the clear and distinct perception of that which I state, which would not indeed suffice to assure me that what I say is true, if it could ever happen that a thing which I conceived so clearly and distinctly could be false; and accordingly it seems to me that already I can establish as a general rule that all things which I perceive very clearly and very distinctly are true.[1] [*Third Meditation.*]

It would be wrong, however, to see in the Cartesian theory of knowledge a simple generalization of the conditions which are revealed in a particularly privileged case. In fact, the experience of the *cogito* was grasped in the movement of its intentional acts, seeing that thought exists only as thought *of being*. Hence, the positing of the thinking self envelops the totality of thinkable reality, and the rule of evidence has no other task than to explicate the very content of the *cogito*. This is what is thrown into relief in the controversy with Gassendi. The empiricist persists in turning the self into a thing and demands of it an 'analysis' similar to a chemical analysis:

> It is not enough to inform us that you are a thing which thinks, doubts, understands, etc., but that you ought to scrutinize yourself, as it were, by a chemical method of procedure in order to be able to reveal and demonstrate to us your internal substance.[2]

To this, Descartes replies by referring to intentionality: the being of thought consists in the very act which posits the constitutive meaning of the object:

> Just as in wax we are able to distinguish many attributes, one that it is white, another that it is hard, a third that it can be liquefied, etc., so also in mind we can recognize as many – one that it had the power of being aware of the whiteness of wax, another that it possess the power of recognizing its hardness, a third of knowing that it can be liquefied, i.e., that it can lose its hardness, etc., when it can be clearly inferred that nothing yields the knowledge of so many attributes as our mind, because as many can be enumerated in its case as there are attributes in everything else owing to the fact that it knows these; and hence, its nature is best known of all.[3]*

But since consciousness is defined in its own particular essence as the possessor of being, it can find the criterion of its value only in its intrinsic evidence, and to seek an external guarantee immediately appears as lacking in meaning. Recourse to divine veracity seems intolerable, not only because it implies a *circle* (seeing that the existence of God has been postulated precisely in the name of a clear and distinct perception), but, in turn, and most importantly, because it pertains to the very *meaning* of evidence to exclude every foundation other than itself. Evidence can be corrected or annulled by other evidence. But such a movement precisely presupposes the absolute right of evidence in general.

Yet, it is worth noting that the guarantee of divine veracity is not some sort of a subsidiary thesis to be introduced from time to time as among the consequences flowing from the existence of God. For the very problem of the existence of God is posited, and rightly so, at the beginning of the *Third Meditation*, solely due to the necessity of founding anew the right of evidence.

> But every time that this preconceived opinion of the sovereign power of a God presents itself to my thought, I am constrained to confess that it is easy to Him, if He wishes it, to cause me to err, even in matter in which I believe myself to have the best evidence . . . . And, certainly, since I have no reason to believe that there is a God who is a deceiver, and as I have not yet satisfied myself that there is a God at all, the reason for doubt which depends on this opinion alone is very slight, and so to speak metaphysical. But in order to be able altogether to remove it, I must inquire whether there is a God as soon as the occasion presents itself; and if I find that there is a God, I must also inquire whether He may be a deceiver; for without a knowledge of these two truths, I do not see that I can ever be certain of anything.[4]

In order to do full justice to the problem, it should be noted that it is not enough to say that since a clear and distinct perception is actualized in an instant, its results will need a new guarantee for other times that it is not an experience that is actually being lived through. For these results are retained in the memory, whose fidelity is not at issue here. And, to posit a truth is to posit it as valid for everyone and for all times. If, then, the claims of evidence are incontestable at the moment of its actualization, they should be understood as faithful transmitted to memory, to the extent that we are careful in this undertaking. But this is precisely what Descartes does not admit.

> When I consider the nature of a triangle, I recognize quite clearly that the three angles are equal to two right angles and it is not possible for me not to believe this so long as I apply my mind to its demonstration; but so soon as I abstain from attending to the proof, although I still recollect having comprehended it, it may easily occur that I come to doubt its truth, if I am ignorant of there being a God.[5] [Fifth Meditation.]

The problem is totally insoluble, if truth is taken as an absolute *in itself* which, once known, could no longer be called into question: the very need of a supplementary guarantee would involve a contradiction in terms. But it was precisely truth, for Descartes, that was not originally understood as an *in itself*. Its whole *being* consisted in *being perceived* clearly and distinctly in the actual movement of subjective life. The problem of divine veracity marks the exact moment of surreptitious passage from the original conception of truth as experienced evidential signification to its commonplace interpretation as an *in itself*.

This can be confirmed, if we recall the conclusion that Descartes gives to the experience of the *cogito:*

So that after having reflected well and carefully examined all things, we must come to the definite conclusion that this proposition: *I am, I exist*, is necessarily true each time that I pronounce it, or that I mentally conceive it.[6]

Thus, the most certain truth of all is not an *in itself* independent of the actuality of the concrete consciousness. The proposition: *I am*, is true only *each time that I pronounce it or conceive it mentally*: not, to be sure, that it becomes false when I no longer conceive it, but its entire existence as truth is solely dependent on the very act by which I know it. Its whole *being* consists in *being constituted* in the temporality of my actual evidence. Whence the limits imposed on the claims of clear and distinct perception are not because of its intrinsic insufficiency but because of the very nature of the object. The value of evidence is effective only at the moment in which it is experienced, simply because the true exists in a true sense only at that very moment, so that there are no grounds for claiming an eternal certification for it. To be sure, the true necessarily is posited as true for everyone and for all times. But it can be a matter only of an actual universality – i.e., of the possibility of an indefinite repetition that is actually experienced. The act which posits the truth as capable of being experienced again and again implies, certainly, an understanding of eternity, but this understanding has its value only in its own ever repeated actualization. Hence, the Cartesian doctrine of evidence should have developed a philosophy of pure immanence which would aim at *reactivating*, at each instant, the *original* evidence in which being is constituted in its *true meaning*. But such concepts did not prepare the way for an understanding of this type. While defining 'eternal truths' as *created natures*, which were readily recognized as encompassing, in truth, the same temporality as the act of knowledge itself, Descartes falls into the popular misconception that takes them as absolute realities: a naiveté which is repudiated immediately by his unexpected demand for a supplementary guarantee and a mystical recourse to divine veracity. Such a *non sequitur* expressed only the self same structure of the constituting act, seeing that it necessarily posits its object independent of the movement which engenders it, and in its turn, grounds the meaning and value of this movement. The Cartesian circle was originally concealed within the essence of truth, since such an essence is *concealed in being revealed*.

18. PHENOMENOLOGICAL DESCRIPTION AS A CRITIQUE OF AUTHENTICITY: STATIC AND GENETIC CONSTITUTION

It is now time to grapple with the major difficulty raised by the very principle of the phenomenological method. How is the pure description of lived experience going to resolve the problems of the critique of values, which, traditionally, has been raised by philosophical reflection? We have, to be sure, singled out [self-] evidence as a privileged experience, which is not a passive given but the *primordial mode* of intentionality, whereby there no longer can be any doubt, since doubt can be meaningful only in terms of it. But have we not renounced, thereby, the search for a strictly philosophical foundation? If in approaching a man, I realize that it was just a dummy, I have corrected one perception by another, which implies the general right of perception to posit its object as a sensible reality. And if I correct an error in calculation, I can do so only by doing another calculation, which does not release me from the type of evidence proper to mathematical reasoning. Apparently, phenomenology would be capable of making evident nothing but procedures of this kind. It is, indeed, "the principle of all principles . . . that whatever present itself in 'intuition' in *primordial form* (as it were in its bodily reality), *is simply to be accepted as it gives itself out to be*."[1] The philosopher, then, would be forced to recognize his own uselessness and to leave to spontaneous consciousness the task of resolving its problems by its own movement.

However, phenomenology always has presented itself as the only valid mode of giving a philosophical critique. The principle of evidence is not a dogmatic theory but a simple starting point that outlines the plan to be followed in achieving this undertaking. Phenomenological description is proposed precisely in order to rectify significations intended by consciousness, by explicating their *authentic content*. And, as a matter of fact, the critique is fruitful only to the extent that it reveals to the subject the profound intentions of his own procedures. So, also, is the very dialectic of spontaneous consciousness whenever it is surpassed and elevated to the level of philosophical reflection. Phenomenology does nothing but take up again, in a methodical manner, the movement by which natural thought puts its own evidence into question at certain *critical moments* by a spontaneous return to the consciousness of self.

Thus, a geometrical problem could not be resolved in any other way than by geometrical reasoning. But at each decisive stage of the scientific development doubts arise in regard to the very nature of the scientific object and of the procedures that establish it. There could be no basis for contesting the value of mathematical thought as such: nevertheless, there should be a critique of it in the form of an explication of the *true meaning* of the fundamental concepts that define the structure of its domain and, correlatively, of the *type of evidence* whereby they are established. Whether one goes back to the discovery of irrational numbers, infinitesimal calculus, imaginary numbers, non-Euclidean geometries, or of whole numbers, we shall see that, in each instance, progress could occur only by reflection on the actual signification of the mathematical object so far as it is *intended authentically* as a *pure possible* that is given in *pure operations as such* in the creative activity of the mathematician. Inasmuch as *mathematical existence* is demonstrated by simple noncontradiction and conversely is precluded by every other principle, the '*problem of existence*' was *forgotten* consistently in the technical work in which the conditions of sensible intuition were confused with operational concepts and, accordingly, their development was limited in an arbitrary manner. For this reason, fractions, finite quantities and Euclidean space for a long time enjoyed the privilege (which was nothing but a prejudice) of actual 'reality'. Each time, then, that the progress of *practical* science demanded an extension of the domain of science – an extension that was opposed, precisely by the habits which, at the preceding stage, had made possible this very practice – there had to be a *return to a primordial consciousness* in which the actually experienced meaning in mathematical thought was revealed once again in absolute evidence, seeing that mathematics defines its object by means of pure constructs that do not imply any other condition than a simple logical coherence and which, therefore, are independent of any sensible intuition. In other words, it is a question of a *critique of mathematical evidence* being brought about in the absolute evidence of self-consciousness by a *phenomenological explication* of the authentic intentions of the mathematician.

The possibility of a critique of this kind rests on the very structure of spontaneous evidence, to the extent that the ontological meaning that is constitutive of the *being* of the object remains, of necessity, implicit in the ingenuousness of its lived experience. Thus, the physicist operates on physical objects (e.g., facts, laws, theories) without explicating the *meaning of the being* (*Seinssinn*) of the physical as such. And so, fundamental concepts (such as space, time, mass, etc.) stand out as so many things *in themselves* independent of all reference to reality as it is lived. The work of the great creators was to dispel such pseudo '[self-] evidence' by a return to the original meaning of notions, in the *movement* by which they are based upon a *physical experience*. Thus, spatio-temporality is not a natural category to be imposed upon one's perception of the real world. It belongs rightly to its own *meaning* to be defined in the *actual* operations of substitution and the establishment of relations – the accomplishment of which varies, obviously, according to the progress made in developing mathematical apparatus and instruments of measurement. Accordingly, space and time, conceived as homogeneous and infinite media, made sense solely in terms of the possibility of obtaining measurements independent of the system of reference with which we work. Such a possibility implies, evidently, that only a simple substitution will suffice to make two different reference determinations coincide. But the condition that is achieved for a system of solids that are not remote from each other – like the world of our ordinary experience – is no longer for reference frames that move with great speed in respect to one another The very notions of space and time must, then, be reconstituted, seeing that the actual conditions of operations of measurement do not represent simple contingent circumstances but the very foundation of theoretical concepts, as the *condition of evidence* that is constitutive of the domain that is being studied. The Einsteinian critique liberated physical thought from its traditional points of view by explicating the *meaning of the being* of the physical domain as the very same one that is based on an actual experiment. It is this very same meaning that the physicist *intends* authentically but which was covered over by the habits of technical work, in which notions are taken for things and detached from the dialectic which engenders them in their actual being and truth. By the same token, the discovery of the 'principle of indeterminacy' depended on an explication of the *meaning* of the very concept of determinism inasmuch as the foreseeability of the position of a body at a given instant implies the *practical* possibility of defining, simultaneously, its position and speed at the previous instant. That such a possibility is limited by the *real* conditions of experimentation entails a corresponding modification of the principle. But here again, it is necessary to recognize that verification is not an external operation that would consist in some kind of extrinsic comparison between the calculations of the theory and the data of experience. It defines the constitutive *evidence* of physical reality in the actual meaning intended by the physicist – i.e., that without which the apparently best established principles are, in fact, nothing but simple *prejudices*.

Thus, positive thought was already in a spontaneous manner pursuing the critique of the ontological meaning of the object of knowledge by *explicating the intentions that were experienced authentically* by the scientist. It would be easy to find analogies within the domains of feeling and action. The pseudo-problems of casuistry find their solution in a return to the lived *primordial meaning* of values and ends – a meaning that is always present, that is obscurely intended in the worst failures, and the disclosure of which directs the will in its effort to set things right again.

Phenomenological description consists in setting forth in a systematic work the philosophy that is inherent to natural consciousness. It will give, thereby, a new status to the positive sciences by disabusing them of the naiveté that prevents them from understanding "their own achievement, inasmuch as the intentionality that produces them remains in them in an implicit state." Such 'sciences' find themselves

> unable to clarify the genuine being-sense of either their domains or the concepts that thematize their domains. Thus they are unable to say (in the true and ultimate sense) what sense belongs to the existent of which they speak or what intentional horizons it presupposes – horizons of which they do not speak but which, nonetheless, contribute to determine the meaning of the question.[2]

Phenomenology must allow these sciences to recover the authenticity that they had for the great creative scientists, who conceived them solely within the unity of a *universal wisdom*. Modern science became fractured as a result of a proliferation of specialized techniques.

> (It) has abandoned the ideal of genuine science that was vitally operative in the sciences from the time of Plato; and in its practice, it has abandoned radicalness of scientific self-responsibility .... European sciences ... have lost their great belief in themselves, in their absolute significance. The modern man of today, unlike the 'modern' man of the Enlightenment, does not behold in science, and in new culture formed by means of science, the self-objectivation of human reason or the universal activity mankind has devised for itself in order to make possible a truly satisfying life, and individual and social life of practical reason .... This great belief, once the substitute for religious belief ... the belief that science leads to wisdom ... had (at least in wide circles) lost its force. Thus men live entirely in a world that has become unintelligible, in which they ask in vain for the wherefore, the sense, which was once so doubtless and accepted by the will.

> Now, however critical and skeptical our attitude toward our scientific culture as it had developed historically we cannot simply abandon it, with no more reason than that we lack an ultimate understanding of it and are unable to manage it by virtue of such an understanding .... If we are not satisfied by the joy of creating a theoretical technique, of contriving theories with which one can do so much that is useful and win the admiration of the world – if we cannot separate genuine humanity and living with radical self-responsibility, and therefore cannot separate scientific self-responsibility from the whole complex of responsibilities belonging to human life as such – then we must place ourselves above this whole life and all this cultural tradition and, by radical sense-investigations, seek for ourselves singly and in common the ultimate possibilities and necessities, on the basis of which we can take our position toward actualities in judging, valuing, and acting .... We must rise above the self-forgetfulness of the theorizer who, in his theoretical producing, devotes himself to the subject-matter, the theories, and the mathods, and accordingly knows nothing of the inwardness of that producing – who lives in producing, but does not have this productive living itself as a theme within his field of vision.[3]

Thus, the descriptive method should not consist in a passive acceptance of the 'given'. It aims at the acquisition of a universal awareness that does away with the prejudices acquired in technical work and reestablishes the constituting subjectivity in its primordial truth. In a word, the explication of lived experience ought to take the form of a *universal constitution* in which being, explicated as *constituted being*, will be defined and justified in its authentic meaning. This task comprises two moments. The meaning of being in its different domains will be taken at first, just as it presents itself and as subjected to a 'static' description that will make precise the structure of each domain and, correlatively, its modes of appearing within consciousness. But inasmuch as this meaning is implied from now on at each moment of the development of subjective syntheses, it will be necessary to question its very being, seeing that it refers to a *primordial creation* that is always present under a *sedimented* form. Meaning is *engendered* constantly in a movement of production that is "implied as *sedimented history* (*sedimentierte Geschichte*) in each constituted intentional unity and its current ways of being given to consciousness."[4] *Static* constitution intended only to analyze a content of signification that already had been posited. *Genetic* constitution defines the conditions of the act that produces it. Static analysis gives a response to problems of structure. The theory of genesis explicates the *ultimate signification* of constituted being – the ultimate foundation of every conceivable truth.

Thus, if we take the instructive example of the material object, we see that static constitution enables us to distinguish rigorously the 'thing' given in antepredicative experience from the 'objective reality' as defined by mathematical physics. Immediate perception is not a 'rhapsody of sensations' but an intentional movement whose articulations can be described with precision. The 'thing' is grasped in sensible intuition with a *meaning* that refers to a *horizon* of possible perceptions, as the perception of 'the same thing'. To see a tree is to be conscious of the possibility of apprehending it from another side, from nearer or from further, of finding it again after having left it, of subjecting it to a series of operations defined in terms of acquired habits. These possibilities present themselves in the form of a horizon that implies, evidently, a certain indeterminacy of content. I know that the object haas a 'backside' that would be presented to me as the 'front' if I were to walk behind it, but I will know this only after I actually have performed the operation. In the movement of perception, the meaning of the object becomes enriched, is confirmed, is corrected or modified, according to the way in which it presents itself. These different possibilities vary, evidently, according to the object that is being considered. They always imply a certain *general style* that constitutes the *meaning of the being* (*Seinssinn*) of the 'thing'. Thus, the possibilities of displacement and manipulation define a spatio-temporal structure – not, of course, in terms of mathematical form but as a horizon that is immediately lived through and realized in its essentially qualitative, subjective, and relative relations. A relativity of this type implies, nevertheless, a perfectly definable style. Other syntheses constitute the moments of substantiality and causality that are presented, to be sure, only as horizons of qualitative relations inferred from the approximation proper to the *life-world*. In short, sense-perception is defined as an original form of evidence whose truth refers solely to the immediate objectivity that is proper to the sensible object. The intervention of the judgment and, more specifically, under the form of mathematical operations, will constitute a new region of being which realizes the Kantian conception of objectivity: reality is presented therein as a collection of determinations that are defined strictly by the concept. But this universality itself has meaning only by means of the experienced syntheses of real consciousness. The ideal object of science is constituted as capable of being understood in an identical manner by everyone: its atemporality as omni-temporality defines *the mode of its temporalization*.

In this way, static constitution makes it possible to delimit diverse modes of being with more precision than does naive eidetic analysis. It gives a foundation to 'regional ontologies' in the constituting subjectivity. However, we have not really gone beyond the pluralism of essences in order to rediscover their unity within a radical justification: description still remains on the plane of the *eidos* taken as a 'given'. But the return to lived experience has laid the foundation for more profound explications. The synthetic structures that have been disengaged from static constitution are not entirely *a priori* forms that, so to speak, would be imposed from outside the concrete givens of the singular subjectivity. Consciousness, taken in its *essence*, refers only to the *meaning* of the being that it intends – a meaning that is experienced immediately by the *ego* as the *horizon of its powers* at each felt and lived instant (*in einzelnen Pulsen des 'Ich kann' und 'Ich tue'*).[5] *The being of intentionality* consists precisely in *giving* this meaning to the object in its production within the actuality of the lived moment (*Sinngebung, Sinnleistung*).

Hence, analysis goes beyond the consideration of syntheses of the modes of appearing, in order to investigate the *origin* of the meaning that dominates them. This appears right from the outset as a ready made 'formation' (*Gebilde*), which is associated with the properly 'given' elements in order to integrate them within the horizon of an object. In the case of sensible perception, the meaning of the being of the 'thing' animates the sensorial givens and makes them appear as simple *aspects* that refer to other aspects of the same '*thing*'. However, such a meaning has no reality in itself: being an intentional formation, it exists only within the act that produces it. But it is a matter here of a *passive genesis* that occurs in the naiveté of an '*anonymous*' life in which the self is not conscious of its creative activity: the product appears from that time as an available acquisition, which reaches consciousness by simple *association* and confers on the *given* a sense of *intentionality*.

It is this passive genesis that projects at every moment the horizon that explains why the world already has a meaning for us. But such a movement is possible only as the *retention* of a primordial act in which the meaning was created by the absolute evidence of an *active genesis*, but which progressively became weaker and so fell, finally, into the '*unconscious*', which is not to say into a phenomenological nothing but simply into a 'limit-mode' (*Grenzmodus*) of consciousness.[6] We note that it is neither entirely a matter of a simple persistance of impressions nor of an association of sensations and images. The meaning constituted in its fullness by a *primordial creation* maintains itself by a *continuous creation* that alone enables us to grasp the object with its objective value. Association recalls to consciousness the meaning that is continuously and passively reproduced in the unconscious by virtue of a horizon of possibilities that is outlined in advance. The world acquires a meaning through the *possible*, which arises as *the truth of its being*. But the possible can exist as such only by means of the creating act of an absolutely primordial evidence, which is perpetuated in intentional becoming as the becoming of one and the same self.

Thanks to the aforesaid passive synthesis[7] (into which the performances of active synthesis also enter) the Ego always has an environment of 'objects'. Even the circumstance that everything affecting me, as a 'developed' ego, is apperceived as an 'object', a substrate of predicates with which I may become acquainted, belongs here, since this is an already familiar goal-form for possible explications as acquaintive explications – explications that would constitute an object as an abiding possession, as something accessible again and again; and this goal-form is understandable in advance as having arisen from a genesis. It itself points back to a 'primal instituting' (*Urstiftung*) of this form. Everything known to us points to an original becoming acquainted; what we call unknown has, nevertheless, a known structural form: the form' object' and, more particularly, the form 'spatial thing', 'cultural object', 'tool' and so forth.[8]

We should note that pointing back to the *origin* is not reducible to considerations of a causal order. Genetic constitution does not search for empirical antecedents as such (which, being 'causes', would have disappeared already) but the primordial act that is *always present* in the actuality of intellectual life. The origin is implied in the actual moment with an *a priori* necessity inasmuch as it is the *very same one* that gives it the meaning of its being. At any time whatsoever, we have not abandoned the dimension of essence. Sensible perception is possible only through the previous creation of the constitutive meaning of the 'thing' because it belongs to its very essence inasmuch as it is the perception *of the 'thing'*. By the same token, the activity of judgment necessarily refers to the primordial act that created the horizon of being as *ideal being*, since every predication implies such an ontological meaning in the intelligible contents that it intends. It is, thus, not a question, strictly speaking, of going back to the contingent circumstances of the psychological development, but of explicating the meaning of the absolute evidence of the *primordial moment*, seeing that it necessarily precedes actual evidence, where the meaning is no longer engendered in the plenitude of its truth, but is presented as an *acquired formation* that is associated immediately with the properly 'given' elements, and makes them 'appear' with an object-meaning already known. Since there is an anteriority, it will be possible to explicate it only in a *history*: but it is a question of a constituting history. The past is taken in its *essential* signification to the extent that it is implied necessarily in the very meaning of the present. It, nevertheless, remains *essentially past*, since it is precisely by virtue of being past that it gives this meaning to the present: it designates *the historicity of the historical act in the being of its possibility*.

The evidence of primordial creation is not, of course, itself an irreducible 'given' that only has to be verified. Transcendental genesis possesses essential articulations that can be defined precisely. Inasmuch as the meaning is grasped here no longer in the unfolding of its implied content, but in the unity of its total being, it will no longer be a question of 'constituting syntheses' in the sense of static phenomenology, but of *fundamental situations* from which a *sense of truth* emerges in the form of a new horizon of being.

Static analysis is guided by the unity of the supposed object. It starts from the unclear manners of givenness and, following the reference made by them as intentional modifications, it strives toward what is clear. *Genetical intentional analysis*, on the other hand, is directed to the whole concrete nexus in which each particular consciousness stands, along with its intentional object, as intentional. Immediately the problem becomes extended to include the other intentional references, those belonging to the *situation* in which, for example, the subject exercising the judicative activity is standing, and to include, therefore, the *immanent unity of the temporality* of the life that has its '*history*' (*Geschichte*) therein . . . .[9]

We can understand now why the constitution of the *Ideen* gave the impression of a return to Kantian constructivism in a form that came dangerously close to psychologism. The analysis was developed therein on a merely *static* plane, and was maintained at each of its moments within the abstraction of a well defined region. Under these conditions, the object could be defined solely in terms of the unfolding of syntheses that made it appear in diverse manners. Its sense of objectivity was reduced, henceforth, to the *a priori* laws of the movement of successive presentations: it was to fall into the absurdity of a philosophy of the *as if*, and the notion of reality was exhausted in its 'equivalent in consciousness' – the world appearing as a dream, albeit well ordered. On the other hand, since the singular lived experiences refer to the object solely by means of the *a priori* forms that dominate them, one could scarcely comprehend how they could qualify as 'transcendentals' precisely in their concrete singularity, and the paradoxical affirmation of a total parallelism between the psychological and the transcendental, apparently, could confirm only the fall into psychologism.

In fact, the notion of a concrete transcendental idealism found its actual justification in *genetic* phenomenology. Constitution no longer intends the form of syntheses of modes of appearance, but the *meaning of being* as such – a meaning that is produced continuously in a *passive genesis* which itself refers to a *first creation* that it does nothing but perpetuate. In the absolute evidence of the creating act, each region is constituted in its primordial truth as a moment within the horizon of total being, always present in its totality and realized in its different modes in the *primordial flow* of constituting life. The structure of intentionality, with the distinction of the *given* and *intended*, is justified now as the result of the movement by which the *intended meaning* is constituted on the foundation of a *given situation*, as its *true meaning*. The intended is revealed therein in the fullness of its being as the *immanent truth for the given*. Such an immanence presents itself as *given* in a primordial evidence whose form of immediacy conceals the movement that engendered it. Hence, there is no longer any basis for separating different regions of being and idealizing them by reducing them to the correlative forms of synthesis. In the *Living Present* of concrete subjectivity, always present to itself in a constant renewal, the particular '*regions*' arise as abstract moments of the total movement, and every consciousness is consciousness of being *itself*.

The all-embracing cogitatum 'of reflection' is the all-embracing life itself, with its openly endless unity and wholeness. Only because it already appears as a total unity can it also be 'contemplated', in the preeminent manner characterizing acts of paying attention and grasping, and be made the theme for an all-embracing cognition.[10]

19. THE CONSTITUTION OF THE FORMAL DOMAIN: LOGIC AND MATHEMATICS

Whatever be the ultimate ambitions of phenomenology, one must recognize, nevertheless, that the Husserlian investigations remain, for the most part, at the stage of *project*. Be that as it may, the problem of knowledge has been treated in a remarkable fashion: its exposition will enable us to make precise the concept of 'constitution'. We shall begin with the most general domain, that of formal logic, which presents the added advantage of being the object of well-known discussions.

We know how the development of the sciences of nature since the Renaissance made the situation of traditional logic intolerable, since it was unwilling to renounce its Aristotelian content and demonstrated again and again its inability to understand the procedures of actual knowledge. On the other hand, the Cartesian effort at liberation, exclusively by presenting evidence for the opposition between the fruitfulness of mathematical thought and the sterility of the theory of the syllogism, did not succeed in giving a rigorous foundation: for whatever be the strength of the intrinsic certainty of mathematics, it nevertheless remains that its necessity, to the extent that it is discursive, necessarily implies an 'analytic' definition and thus refers immediately to the principles of logic. The reconciliation that was attempted by Leibniz did not reach any actual realization, and Kant took note of the failure by postulating, in addition to the analytic necessity proper to logic, a synthetic *a priori* necessity. But this was simply a case of baptizing the problem: and yet the fact remains that mathematical procedures were presented under the appearance of a specific originality. It was the task of the English school of the nineteenth century to elaborate the practical conditions for an actual solution by giving a mathematical form to the content of logic. On the other hand, the progress of formalization tended to present mathematical deduction as a purely logical process. Thus, unity was reestablished in the technical work itself: it became all the more urgent to take this problem up once again on a philosophical plane. For the formal homogeneity of the procedures did not overcome the fact that their usages in every case obeyed radically different intentions: logic and mathematics incontestably correspond to two different vocations, and even though their object is undoubtedly the same, we are aware that it cannot be taken in the same sense. Of course, the problem could not be resolved by an abstract discussion. The difficulties will be cleared up only by explication of the *meaning* of the being considered, just as it is revealed within the evidence where it is *constituted*.

Logic is commonly defined as the study of the laws of veridical thought. This is a defective formula that almost unavoidably leads to a lapse into psychologism. In fact, the logician does not concern himself with subjective acts of thought but with their objectively constituted products – i.e., propositions, their moments, and their combinations. The normative laws that he is able to enunciate are but practical applications of objective laws that regulate propositions and are of a strictly logical nature. Thus, the principle of non-contradiction expresses, initially and in an objective manner, that "of two contradictory propositions, one is of necessity false," from which we infer the rule that one must not affirm and deny at one and the same time the same predicate of the same subject.

Thus, the domain proper to logic is defined initially on the plane of the object as the set of propositions with their various moments and possible combinations – in other words, as the set of *significations* that can be expressed in discourse. It will, accordingly, be suitable to take them, at first, as 'simple significations' by abstracting from every problem of truth or coherence. Thus, we shall have delimited a fundamental stratum whose form, it is easy to see, is being subjected to a certain number of laws – those that allow the signification to exist as such. Thus, when I say: "*the circle is square*," I express a self-contradictory proposition, but one that remains a proposition, since I have given it a predicate – albeit, one that is self-contradictory. But if I say: "*and is square*," I have enunciated a simple string of meaningless words, '*and*' being a conjunction that could not serve the function of subject. It is clear that the conditions for constituting a verbal combination into a meaningful statement are independent of the nature of the objects that may enter into the judgment. Whatever be the subject matter of discourse, the grammatical subject will have to be a substantive (or a nominalized expression exercising the function of a substantive), and the predicate will have to be an adjective or an expression serving the same function or at least some such related word. By considering subject and predicate as pure indeterminate variables, but in every instance maintained as identical with themselves, we can construct a *pure doctrine of the forms of signification* or a *purely logical grammar*, which will establish the formal laws for meaningful propositions and the procedures for the construction of their possible forms. Thus, by taking the predicative judgment '*S is p*' as the fundamental element of discourse we will construct '*S is non-p*'. By nominalizing it, '*Sp is q, S non-p is q*', etc. According to the form of implication, '*if A then B*', we obtain by nominalization: '*if* (*if A, then B*), *then C*', etc.

The pure doctrine of forms of signification defines the conditions of *vague comprehension*. When we speak or listen in a distracted way, we often let pass that which is incoherent or absurd, since we are conscious of grasping or expressing meaningful propositions whenever discourse obeys at least the rules of pure grammar. But we can, equally well, pay attention and force ourselves to understand *in a distinct manner* the statements that are being discussed. In this case we perform the predicative operations in all of their articulations. It is then revealed that such a proposition in being accepted as meaningful cannot be enunciated in a clear manner. One can say, absent-mindedly, '*Sp is non-p*'. But it is impossible, in fact, 'to make' such a statement 'intelligible'. For within a distinct idea, *Sp* excludes *non-p*.

Thus, there appears a new stratum, formed of actually realizable significations and capable of being presented within *the evidence of the distinction*. If we proceed to a *free variation*, and thus eliminate all the material content of discourse, the classical principles of identity, non-contradiction, and excluded middle will remain as *invariants* that are taken simply as the conditions of the pure internal coherence of statements and without any reference to a possible truth as corresponding to the *thing itself*. Thus, the set of actually realizable and '*distinct*' propositions is defined in terms of its form by a *pure analytic of non-contradiction* or an *analytic of pure consequence*.

The evidence of the distinction guarantees merely the simple coherence of the statement; its truth is given in the *evidence of clarity* where the object presents itself exactly as the judgment posited it. It is evident, *a priori*, that a proposition can be verified in experience only if it is already actually realizable in thought. The conditions of the evidence of the distinction are thus necessarily valid for the evidence of clarity. In other words, the analytic of non-contradiction immediately defines the content of a *formal analytic of truth*. But the two levels are clearly distinguishable even though the technical content remains the same: the consideration of a correspondence with reality in the evidence of clarity implies in the very form of the judgment a signification that is excluded *a priori* from the pure analytic of non-contradiction – i.e., that the logical principles determine not only the form of every *distinct judgment* but also the form of every *possible truth*.

We now understand why traditional logic remained disconcertingly sterile despite the fact that its authentic content had been admired for centuries. The three strata that we have distinguished – the doctrine of the forms of signification, the analytic of non-contradiction, and the analytic of truth – were jumbled in a confused way for the classical logician who was working, in fact, solely with pure significations: judgments as 'simple judgments' turned his attention to the formal conditions of a possible truth, in the sense of a correspondence with things. It then became practically impossible for him to thematize the domain of significations as such and to develop for himself the system of forms of possible propositions as pure significations as such. He thus necessarily restricted himself to elementary remarks and to certain trivial applications that imposed themselves by their evidence but were discouraging because of their impoverishment. In fact, the systematic study of significations as such required a special act of objectivation – i.e., the nominalization that transforms the predicative content into an object and makes possible a new syntax. Thus, the collective judgment '*a, b, and c are p*' is sterile in itself. The nominalization of the conjunction '*a, b, and c*' gives the 'set (*a, b, c*),' which '*is p*,' and provides, as such, matter for new operations of conjunction, disjunction, etc. By the same token, the judgment of relation '*a is equal to b*' is workable only through nominalization, and thus it is an available product – i.e., the equality ($a = b$). The judgment of numeration '*an A and an A are p*' gives by nominalization the number '2*A*' which 'is *p*'. This is precisely the form by which the mathematician understands the domain of significations. One could believe at first blush that mathematics is concerned solely with forms of the *object*: sets, relations, numbers, etc. In this way, mathematics would be radically opposed to itself, in terms of a formal doctrine of being or formal ontology and in terms of logic, defined as the formal doctrine of propositions or *formal apophantics* (*apophansis*: predication). But, in fact, it always has been a matter of merely predicative forms that are sometimes grasped at the inchoate stage in the judgment and at other times objectivized by nominalization. Number is constituted in the judgment of numeration; the set, in the collective judgment; and relation, in the judgment of relation. But inasmuch as these contents appear solely in the judgments themselves, they are not manageable. In fact, they are still only 'preconstituted'. Their 'constitution' is realized fully only by means of the *nominalization* that transforms them into independent *objects*: sets, numbers, and relations which can become the basis for new judgments.[1]

We see immediately why mathematics because of its fruitfulness seemed to be opposed to logic, while its procedures were characterized by a formal necessity that could be defined only as analytical, and hence logical. Every judgment has a bearing on an object. To speak is to speak *of* something – to express a signification in regard to this thing. Hence, the signification so designated is not an *object*: it is still only experienced in the *noematic sense*. If I say that "*the garden has flowers*," the object about which I am speaking is the garden itself about which I express a 'state of affairs' (*Sachverhalt*) – namely, that *it has flowers*. But this 'state of affairs' is presented until now solely as intended meaning in the act of judgment: it is not the object itself. The logician reflect on this meaning, the judgment as 'simple judgment', and immediately experiences the impossibility of really performing the predicative operation whenever the meaning entails a contradiction. By freely varying the material content of propositions, he will come to an *invariant* that expresses the formal conditions of all real meaning according to necessary principles: but at the same time he finds himself incapable of going beyond a few common-sense remarks, since his reflection bears only on the intended meaning as it is intended within lived intentionality. He thus ascertains through evidence and an absolute generality that once a predicate has been posited it could not possibly be denied and once denied it could not possibly be posited. He expresses this by stating that of two contradictory propositions one is necessarily false. But he is incapable of constructing in a systematic way all the forms of possible propositions. The difficulty is particularly noticeable for the judgments of numeration or relation, which imply an infinite diversity of distinct forms; its thematization is not possible in the simple reflection on the intended meaning as such. On his side, the mathematician takes this meaning in its nominalized form, *qua* object. He is concerned with sets *qua* sets, numbers *qua* numbers, relations *qua* relations. He, thus, can work them into new syntheses, compose and decompose them *ad infinitum*: but he will always be dealing with objects and not judgments. However, it is only a matter of *categorial* objects whose whole being consists in being posited in a judgment (*kategorein*: to predicate). The truth of mathematics consists precisely in its not having any other concern than that which is constituted in predicative syntheses.[2] But the *intention* of the mathematician does not extend to these syntheses as such. He experiences the necessity of the result without being able to explicate its foundation: for he has abstracted from it in his work as mathematician.

Thus, it turns out that logic and mathematics, though reciprocally ignoring each other, were concerned with the same domain. The logician studies judgments, and the mathematician, the intelligible products that are constituted therein. If I say that '*a, b, and c and p*', I state a collective judgment in regard to the objects *a, b*, and *c* (as perceived individually), and these are brought together again, only by the conjunction '*and*'. If I nominalize, I obtain the *collection itself* as the intelligible (categorial) determination of the object, since from here on I perceive a, *b*, and *c* no longer as discrete objects that have to be reassembled by a conjunctive synthesis but as already a unique collection *which is p*. But it is apparent that here there is only one self-same object with which we are dealing: namely, the predicative contents that are thematized sometimes in a noematic reflection and are nominalized at other times in an objective study. Once the relation between these two disciplines has been made clear, a possibility seems to appear for a collaborative effort in which logic contributes the certainty of its principles and mathematics the fecundity of its development. Mathematical deduction would lose all real meaning, were it reduced to a pure game of empty symbols. In fact, the axioms with which it begins do not define simple mechanical rules for the blind handling of signs. They imply actual operations of judgment and thus necessarily refer to logical notions. Thus, in the formalization of addition, one must not say simply, for example, that "One is always permitted to manipulate the given signs in such a manner that the sign '$b + a$' can always be substituted for '$a + b$'," but that "there shall obtain among the *objects* belonging to the domain (conceived at first as a simple 'something' – as pure objects of thought) a certain *combination-form* with the formal law: '$a + b = b + a$', where the word 'equality' has the precise sense of actual equality, such as belongs to the categorial logical forms."[3] And it is indeed clear that an application of mathematics to reality would not be conceivable if the sign '=' did not have the precise meaning of the relation posited in actual judgments of equalization.

Whatever might be the intrinsic evidence of these results, the problem does not seem to be resolved completely. For once it is admitted that logic and mathematics represent two ways of studying one and the same domain (namely, the ideal domain of categorial forms that are constituted in predication), it would seem that we still remain on the plane of judgment as 'simple judgment': we have not grounded its relation to the object in an effective manner inasmuch as the forms of discourse must define the conditions of *possible truth*. One can, of course, distinguish, on the technical plane, forms of signification and forms of object; '*S is p*' being nominalized into '*Sp*'. But the forms of the object under discussion (e.g., the state of affairs, substrate, property, collection, number, relation) still concern only 'whatever object', (i.e. the 'something in general') that is nowhere given if not in the formal statements themselves as an indeterminate variable. The 'formal ontology', in the way that we have defined it until now, has not yet acquired its true meaning as the formal doctrine of *being*, understood, to be sure, as *real being*. In short, our explications have not yet gone beyond the purely static level of the syntheses of modes of appearance. We have described the manner in which the formal domain presents itself to the logician and the mathematician respectively. We have not yet subjected to question the *very meaning of its being* – a meaning that is always presupposed at every moment of the operations that constitute it. Hence, obscurities reappear as soon as we go beyond the level of methodology, in order to approach the problem of knowledge. A solution is possible here only by a *return to the origin* that explicates the movement by which the domain has been brought about.

The judgment as 'simple judgment' results from an abstraction performed on spontaneous thought, which is always thought *of the object* in the sense of being a real object. To judge is to say something *of* something. I say of the garden that it has flowers. The proposition '*the garden has flowers*' is not a 'simple proposition' in me: it posits a *state of affairs* that determines reality itself. Undoubtedly, this state of affairs is not yet presented as an objective determination: it is only the *meaning* through which the subject grasps the thing of which he speaks: namely the garden itself. But it suffices to nominalize this meaning in order to insert it into the object as a real determination. In the new proposition: '*that the garden has flowers is beautiful*' or again '*the flowered garden is beautiful*', that of which I speak is the garden with the determination of having flowers, and of this object I say that *it is beautiful*. But the state of affairs thus objectivized is recognized as being the same as that which was simply posited in the lived act of predication in the noematic sense: 'the flowered garden' bears an objective determination that is precisely only the state of affairs in the first judgment: '*the garden has flowers*'. And so we see that the judgment is not originally a 'simple judgment' in the subject, but an act that determines reality itself – the determination being explicated as such in nominalization. Thus, science as a system of judgments about reality will be at the same time the system of determination of reality itself: it posits the *concept of nature*, not in the sense in which this concept would be simply in us and not in things but in the sense that it defines *nature itself in its intelligible determinations*. Hence, the formal conditions of truth that the logician proposes to explicate are not to be understood originally as the conditions of judgment but rather as the conditions of reality itself. Thus, logic as the formal theory of science implies initially the meaning of a *formal ontology* or formal doctrine of being.[4]

But the actual work of realization inevitably takes the form of a critique of judgments, understood as 'simple judgments' in us, of which it is a matter of knowing whether they can be applied legitimately to the object. In the experience of error we discover that what we took for a determination of the real was, in fact, but a 'simple judgment' in which the intended meaning was no more than a purely subjective pretension. Science presents itself as an effort to criticize spontaneous judgments, to which it attributes the mere value of being 'simple judgments' and in order to establish 'true' judgments in the sense of being 'scientifically verifiable'. Logic, as the doctrine of science henceforth will be the study of the formal conditions without which a judgment could not possess legitimate claim to truth. In this way it deviates in a sense from its original vocation and presents itself not as a formal theory of *being* but as a formal theory of *judgment* as 'simple judgment' or *formal apophantic.*[5]

This is precisely the work of Aristotle, which appears as the first realization of logic and at the same time as its first *alienation* from its authentic meaning. The development of mathematical formalization completes the movement. The critique of the forms of judgment in their claim to a possible truth implies, as an abstract stratum, a *pure analytic of meanings as such*, which systematicaly constructs all possible predicative forms on the purely operative level. Yet, this is precisely the task that is proposed by traditional mathematics, which thematizes its objects as pure operative products for which no other problem is raised than that of a simple possible construction in *distinct evidence*. We are, indeed, within the domain of predicative significations, but taken as pure significations and absolutely deprived of their original relation to the object, since mathematical 'existence' is defined precisely in terms of the radical elimination of every claim to real existence. Such an *alienation* made possible the technical success of the enterprise. By remaining on the plane of a *pure analytic of non-contradiction*, mathematics has developed systematically the study of predicative forms nominalized into forms of object (e.g. relations, collections) but implying at the same time the forms of judgment wherein they are constituted. There remains only the further task of reestablishing their properly logical significations, as forms of all possible truth. Thus, the results of the *analytic of non-contradiction* (the mathematics of the mathematician) will be transferred to logic as the *analytic of truth*. It is only in logic that the formal system takes on its real meaning as the system of forms of true judgment or formal *apophantic* and as the system of forms of being or or *formal ontology*.[6]

We understand now why the results of static analysis seemed to us to be insufficient. The logical domain was taken as a given upon which it was sufficient to contruct the edifice methodically by going from the simple to the complex – e.g., the doctrine of the forms of signification, the analytic of non-contradiction, the analytic of truth. But if one readily sees the unity of the first two moments, the passage to the third raises difficulties: how can the conditions for the realization of judgment, in *distinct* evidence, define the norms of reality itself that are given in evidence of *clarity*? Despite appearances, the objective direction of mathematical research was not enough to ground the notion of a formal ontology. For the mathematical object is nothing more than the meaning that is constituted in the judgment as such. The return to the origin provided a basis for overcoming confusions. The real order does not proceed from the purely verbal actualization of predicative forms to the consideration of their possible truth. One begins, in fact, with actual judgments in which the world is constituted in its intelligible determinations, and logic is defined primordially as the doctrine of *being*. The notion of judgment as 'simple judgment' is only a *means*, albeit necessary, to criticize the products of spontaneous predication and to accede to scientific knowledge. Thus, the analytic of meanings as such (analytic of non-contradiction) precedes the analytic of truth only on the purely technical plane. In fact, the analytic of truth is posterior to the analytic of meaning, and represents but an abstract moment whose true signification refers back to the actual judgment as the determination of the real. The problem of the passage of distinct evidence to clear evidence or, in common parlance, the problem of the passage of the 'formal' signification to the 'real' signification of the logical, does not arise: *distinction* is but an artificially detached moment of *clarity*. And one also sees that the relation of mathematics to logic does not consist in a simple technical difference in the direction of the thematizing viewpoint. It is a question of the opposition between a purely abstract conception of predicative forms and their full comprehension as forms of being. Thus, logic constitutes the true ontology, while mathematics reaches the domain of apophantic meanings that have been reduced to their purely operative content. But the activity of the logician, performed in the form of a critique, is developed traditionally within an apophantic. Mathematics, on the contrary, objectivized significations in order to handle them better and thus placed itself in its surface appearance on the level of an '*ontology*'.

We will note that in this genesis of the formal domain we have considered not the empirical antecedents of logical thought but the authentic origin of its *meaning* inasmuch as it necessarily refers back to the concrete movement of actual judgments. The passage of spontaneous thought as the dogmatic position of the determinations of the real object to the critical consideration of judgments as 'simple judgments' and to the examination of their formal structure does not constitute a contingent circumstance of the discovery of the logical domain: for it is precisely the implicit presence of these actual predications that gives to the region of pure significations as pure assertions of discourse its authentic meaning. Thus, the anteriority of real judgments is imposed in a necessary manner, and we are compelled to accept the concept of a history as a *constituting history*, even though it is taken in its intelligible content only. But then, the problem of its relations with actual history arises. Nevertheless, before taking up that problem, we still have to specify the meaning of the logical domain. To the extent that the judgment exists fully and truly as a real judgment, the forms that define it on the purely categorial plane as pure predicative operations are no longer separable from the material content that it implies. From the predicative evidence, we are referred back to the *antepredicative* as the condition of the rationality of Logos.

When we say that logical laws, taken in their fullest sense, define the formal conditions of being, it is not a matter from the very first of sensible reality as such but of its *intelligible being* just as it is determined in predication. Thus, the principle of identity signifies only that every 'state of afffairs' that is affirmed of the object is valid as such for all times and places: for the predicative content exists only as ideality. If I write on a round table, the roundness is a real moment that is constituted in sensible syntheses and, more specifically, kinaesthetic and oculo-motor syntheses. The shape is immanent to the object itself; it changes and disappears with it. But if I say: "*the table is round*," I posit a state of affairs that, as an ideal determination, will always remain identical to itself. The shape of the table, as the table itself, exists only in the moving flow of sensible presentations. The fact that "*the table is round*" is a categorial (predicative) truth that is valid as such for all times and places. Undoubtedly, it is a matter of an occasional judgment that refers to the *situation* in which I find myself. But its content as such is defined as a pure ideality whose being is beyond space and time. If I say: "*The Trojan War has taken place*," I refer to a past event by positing an intelligible content that is not, as such, in time, but belongs to the ideal domain of predicative significations. The principle of identity has no other meaning than the defining of this ideality.

The principles of non-contradiction and excluded middle should be interpreted in the same manner. An ideality *qua* ideality always implies its possible negation. Its being true consists precisely in excluding this possibility of negation and vice versa. The perception of the roundness of the table does not contain within itself the exclusion of non-roundness, for there is nothing in roundness as a sensible shape that could cause the possibility of a shape that would not be round. The table is simply perceived as it is. Corrections are, undoubtedly, possible. Seen more closely, the object can reveal itself to be oval. But the 'reversal' can arise only with the very experience of 'deception': its possibility is not anticipated in the original perception. But the state of affairs (i.e., '*the table is round*') defines an ideal content, which in itself can be the object of a negation as well as an affirmation. The meaning of the one consists in the exclusion of the other, and the exclusion of the first reverts to the positing of the second: *tertium non datur*.

We should note, however, that if the predicative contents by themselves constitute pure idealities, it belongs to their actual meaning to determine the object *itself*. By saying "*the table is round*," I am obviously speaking of the real table that I have before my eyes, and I am enunciating only a 'state of affairs' that belongs properly to it. When the physicist constructs a theory in the form of a system of equations, he does not place himself in an ideal world that is other than that of sensible perception: he, obviously, intends no other world than *this very world* in which we all live and which science defines in its true being. Thus, the idealities constituted in predication could not very well be separated from sensible reality as such: they are, precisely, none other than the intelligible determinations *of* this reality. Thus, logical principles that define the being of idealities cannot be, in their profound sense, completely independent of the material content of actual propositions. For their truth consists precisely in being true *of* this content.

The illusion of formalism is based on the very conditions of the task of formalization. The formal *a priori* presents this peculiar characteristic of being obtained in an absolutely limitless variation, an 'evacuation' (*Entleerung*) of all content. If I want to obtain the essence of a determinate region (the 'thing', animality, society, etc.), I must preserve the material meaning that is proper to it by intuiting concrete examples in which I experience the necessity of its *a priori* content. Thus, on the level of a purely blind or symbolic thought, I could conceive of a 'thing' that would not be extended. But if I try to realize this in imagination, I immediately become aware of the fact that spatiality belongs to it essentially. Yet, formalization is presented under an original aspect, since it is no longer a question of an object of a particular region but of an object in general, variation should not be limited by any condition that could be represented in the imagination. It concerns an absolutely indeterminate *X*. The logician thus gets the impression that the laws that he defines are radically independent of every reference to the concrete content of the judgment. If we designate the moments that subsist after a proposition has been emptied of all its properly predicate relations and that irreducibly refer back to an antepredicative intuition as '*kernel*', we shall assert that in the formal domain that has been thematized by logic the *kernels are irrelevant*.

But that is only so on the surface, for it is quite evident that logical laws concerning actual judgments could not be entirely unrelated to their content. By substituting for imaginative variation, properly speaking, the purely formal consideration of an absolutely indeterminte *X*, we have made use of a necessary technical means, but one that conceals the authentic meaning of the operation. The content is indeterminate in the sense that it can be dealt with in any region whatsoever: yet, it is necessary that such a region exist, and its existence will imply, necessarily, conditions of a material nature. In other words, the indeterminate *X* does not designate anything whatsoever, but a possible reality, whatever such might be. Thus, logical laws could not be applied to a proposition such as '*the sum of the angles of a triangle is equal to the color red*'. For this statement, while respecting grammatical forms, has no imaginable meaning: it does not belong to any 'region' of being. We could not say that a proposition of this kind is false: for it would follow according to the principle of excluded middle that its contradictory proposition would be true. And so, it is neither true nor false but simply 'devoid of meaning'. Thus, logical laws are only formal when they are considered abstractly. Their truth is related to the material conditions of the true judgment: *the kernels are relevant.*[7]

20. THE GENESIS OF JUDGMENT

Reflection on the formal domain has permitted us to go beyond formalism. Form has no other meaning than as the form of being: it is, thus, a moment of being itself. And even though it first of all concerns the ideal being of intelligible determinations, it turns out that the idealities that are constituted in the judgment refer irreducibly to the realities that are presented in sensible perception. Thus, the truth of predicative forms is founded on the movement of antepredicative experience. Logic, as the theory of the forms of intelligible being, becomes fully justified only by explicating the dialectic by which the intelligible is *engendered* from the sensible. This was precisely the pervasive direction of Husserl's thinking as is already attested within the sixth *Logical Investigation*. But the properly genetic theme will be explicated only by 1919 in a series of unpublished manuscripts, the essence of which has been presented by Landgrebe in *Erfahrung und Urteil.*[1]

We immediately begin with the antepredicative world, in which 'things' are given as spatio-temporal unities in possession of diverse qualities that are integrated in a collection of habits that makes them appear to belong to more or less familiar *types*. An organization of this kind is immediately evident only in appearance. Even though it still has nothing of the precision and rigor of the categories of the understanding, it nevertheless implies a determinate style that refer to an extraordinarily complex *genesis*, to which we will have occasion to return. We provisionally understand the *life-world* as a given in order to consider the movement that elevates it to the forms of intelligibility.

Sensible perception takes place in a series of syntheses of *simple identification*, seeing that each presentation is linked to the preceding one as the presentation of *the same thing*. The relation is guaranteed by the horizon in which the course of experience takes place, seeing that each 'aspect' is grasped with a certain meaning, where practical possibilities of development that are relatively indeterminate in their content are sketched out, but which, in virtue of their general style, refer consistently to one and the same object. So, each moment appears as the continuation of the preceding one that *still* remains *present* in retention. In the series of presentations, the perception of the thing becomes enriched by realizing its possibilities; it also can be modalized and so the object will be revealed as different from what it was thought to be. Thus, I discover that this sphere has been dented. The synthesis of identification is still taking place since we always are aware of dealing with the same object. But certain determinations stand out as being nonexistent and motivating a correction of the preceding perception. Thus the spontaneous course of sensible experience already implies, in the case of modalization, a movement of antepredicative abstraction in which moments (forms, colors, etc.) are separated from the object that is retained as a unified whole. And so an effort of attention makes possible the explication of certain determinations while maintaining them as existing within the object: this is the process of 'explication'.

Let us look at this particular cup. We look it over and concentrate on its roundness and then an aspect of this shape, etc. In these partial perceptions the cup always remains before us as the constant theme of our attention. But the perception of it has undergone a remarkable modification: it no longer takes place as a primordial act, but as a retentional act, being maintained only in the mode of the '*still*' [*encore*]. In other words, the total object is always present but left in the background, while the particular moments that 'cover it up' precisely in virtue of being moments are explicated in progressive fashion. "At each step of the 'explication' the singular moment which is being explicated covers up the substrate and is incorporated in this way within its significational content"[2]: It appears as a determination of this substrate.

We note that this whole process is accomplished on the antepredicative plane of receptivity.

In this synthesis of the passage of S to p, the substrate that is maintained as 'still grasped' (*noch im Griff*) has enriched its meaning. But while keeping S before our gaze, we pass on to its moment p as 'covering over' S and S as 'being enclosed' within p; we have not yet posited S as the *subject* of a predicative judgment and as possessing its moment *qua* predicate. This is the *result of a new mode of activity*. In a receptive grasp and in its 'explanation', the activity existed already: by being applied actively to the substrate S, the self grasped it in its indistinct unity and thematized it, and then actively grasped its determination p in the explanatory synthesis. Thereupon, the covering up of the substrate S that was maintained within the gaze of consciousness in the mode of the 'still' by its determination p, was inserted *passively* as a result of the 'explanation'.[3]

Thus, on the level of antepredicative perception, the *activity* of 'explanation' has brought to light the relation between substrate and determination as a *passive* form. If instead of considering this cube in its immediately perceived unity, I concentrate especially on its red color, the cube is *still grasped*, but simply as the *background* upon which the particular moment of color or form *qua* figure stands out: the relation appears as ***passively lived and felt***, but its presence will occasion an ***active repossession*** that will constitue the task proper to predication.

The substrate S, whose meaning is enriched in the course of the process, is left in the background and is maintained only in the mode of the 'still grasped'; it is no longer the center of attention, the 'focus' of the gaze of consciousness. But being presented now with its enriched meaning, it returns to the center of attention. We return to S, identifying it thus with itself; i.e., it comes once more to the fore by virtue of being S: in this new thematic grasp, we have its enriched meaning as simple protention in connection with the retention of the process that has just elapsed [i.e., 'the process of explanation'] * Attention is actively directed toward S taken in its enriched meaning, which implies that we return *anew to p*. For, at first, p, as enrichment of meaning, emerges in only the synthesis of the passage in which it is explicated as a determination that 'covers up' the substrate. But the passage is directed now by the will to know, which aims at positing S in a clear way in its determination. An *active* intention tends to grasp explicitly what was previously but a *passive* covering up; in other words, to effect actively the passage to p and, in a primordial act, to create this enriched meaning for S. As an active self, directed to S in its enriched meaning and focusing my attention on this very enrichment, I achieve the passage and partial covering up [i.e., the determination p being posited as a moment that covers up a 'part' of the total object S] as a free activity, and I thus fulfill the determining intention – the intention that aims at S in its enriched meaning by the passage and covering up. I have S as the substrate of a determination and actively determine it. The object that serves as substrate takes the form of predicative subject. It is the theme of which we speak, as the *terminus a quo*: and the activity bears upon the predicate as the opposite pole, the *terminus ad quem*. In an activity that is not merely a synthetic activity in general but the *activity of synthesis itself*, we become conscious only now that S is determined by p in the mode of '*S is p*'.

. . . That the substrate of the 'explanation' becomes the subject and the determinations explicated simply at the level of receptivity become predicates can be realized only inasmuch as one's attention returns to the somewhat hidden unity, which in the activity being exercised on the level of receptivity is preconstituted passively on the level of receptivity. *To cling to this unity in order to grasp it is to repeat the process in a modified attitude* – to transform the passive synthesis into an active synthesis . . . . From the spontaneous grasp of S as subject, the act of determining identification bears on p. Previously, already, in the 'explanation', the object is determined implicitly as p, more precisely it is revealed as such, but the fact of 'being determined as' was not explicitly grasped. It is only so in an active repetition of the synthesis – a repetition that presupposes a previous 'explanation' at the level of receptivity . . . .

In a word, *the predicative synthesis in essence has always two strata:*

---

* (All instances in which square brackets are used within a quotation of another author are used to indicate that what is found within these brackets are the words of Trân Duc Thao, rather than those of the author he is quoting. – Trans.)

(1) The passage of S to the moments $p, q \ldots$ which emerge by covering up S; p, q are grasped in their own right. The attention that was given to the object-meaning of the preconstitution and, correlatively, to the content that came to light therein as the content of object S, now slips in regard to its determinations. But S, as well as its moments already grasped, remains 'still grasped'.

(2) A new operation is performed on the latter: the attention of the ego returns to S and by once more focusing on it, becomes conscious of the enriched meaning of S and so is 'gratified' by recreating this enrichment by means of a primordial act through which the passage to p is achieved once more.[4]

We have insisted on this description in view of its decisive importance. As strange as the notion of a *passive preconstitution* of ideal significations may appear, it nevertheless contitutes, in the *antepredicative activity* that develops on the level of sensible perception as such, an absolutely self-evident *phenomenological given* and contains the only real solution of the passage from the sensible to the intelligible. The *work of subjectivity* in the *life-world* gives the idealities of the *world of the spirit* their only possible justification, seeing that the actual truth-meaning of their ideal being can consist only in being posited as the *very truth of this world*.

That such a positing implies a specific act does not entail that this specificity indicates an absolute separation; as separation, it is nothing but the actual realization in which the authentic meaning of the unity is achieved.

Each enrichment of the logical meaning presupposes a previous enrichment of receptivity. The spontaneous (predicative) synthesis of identification can take place only where perception and 'explanation' have already occurred at the level of receptivity.[5]

Thus, the creative act of judgment consists only in again effecting in an explicit manner a relation that originally appeared in a passive form in the very movement of sensible activity. Its whole *truth* consists in such an ever renewed *repossession* in which the real is constituted in its *true being*. It was precisely the merit of Kant to have shown that sensation alone allows one to give to the category its real signification: the *transcendental Analytic* was preceded by an *Aesthetic* that defined the conditions by which the categories of the understanding acquired cognitive value. To be sure, the consequence of this was that the constituting consciousness originally is presented as a sensible consciousness, and the intelligible emerges from the sensible itself as its meaning of truth. But the critical philosophy immediately provided itself with the result of the movement. Directly placing itself at the level of universal forms, it denied to itself the understanding of the *mediations* in which they are constituted. Hence, the determining ego is detached abstractly from its sensible foundation without being able to understand how it can be self-identical with the empirical ego in the experience of actual consciousness:

> but the question is to know how the ego (the 'I think') is distinct from the ego which intuits itself while forming with the latter but one and the same subject, how I can say, consequently: *I*, as an intelligent and *thinking* subject, know myself in the same way as an object that has been thought, so far as I am given to myself in intuition, solely as I know other phenomena, that is, not as I am for the understanding but as I appear to myself; this question raises neither more nor less difficulty than to know how, generally speaking, I can myself be an object.[6*]

In fact, the real problem does not concern the relation between the *determinant* and the *determined*: for there is no difficulty regarding the fact that the ego recognizes itself in the very affecting by which it is itself affected in its internal meaning. But empirical consciousness implies a sensorial content that cannot be explained simply by means of the manner in which the ego appears to itself. And since the identity of the two *egos* is unquestionable, one must admit that the act of constituting is homogeneous, from the outset, with the sensible given that it is going to raise to the plane of intelligibility. The transcendental ego, thus, is conceived as temporal, seeing that the act by which it transcends time defines only the supreme form of its temporalization: "*the atemporality of the objects of the understanding*, their existence 'everywhere and nowhere' reveals itself as a particular form of temporality . . . . This *supratemporality* signifies *omnitemporality* . . . . *Omnitemporality is itself a mode of temporality*."[7]

Such is the horizon in which the analyses of genetic phenomenology are carried on. Intentional genesis is placed neither on the abstract plane of the 'conditions of possibility' nor on the psychological plane of mental constructs. The intelligible is not imposed on the sensible in the sense that the sensible from now on would have already been informed in the synthesis of intuitions; neither is it derived from the sensible as the product of a simple combination. It implies a *conscious grasp* in which the *passively preconstituted significations* are explicated in the *real sensible work* that develops in the *life-world*. In the dialectic of this passage, the ego realises itself in its *liberty*, by affirming the truth that it finds imposed on itself in the *passivity* of its living experience. In the movement of its lived experience, all *praxis* brings about a new *horizon passively* in which the being of the real is revealed in its *true meaning*. If I grope in the dark, after a certain length of time I begin to feel certain relations being outlined *passively* in me; and as soon as I become *actively* conscious of them, they allow me to come forth from my immediate situation and to dominate the space in which I have been groping.

Such a consciously becoming aware, however, is not an arbitrary act: it is required by the very presence of these relations that I experience within myself in a more and more pressing manner, to the extent that the groping movements settle in *retention* and accumulate in the form of *global possibilities* that are felt and lived and that are imposed upon me as the *actual meaning* of my real activity. The movements that I have just gropingly made remain *still present* in their general form and *continue to be outlined as practically experienced possibilities* in the current activity; and the ensemble of these possibilities is affirmed within an accumulating evidence as the *horizon of my capacities*, in which the *meaning of the being* of the world that surrounds me is revealed. In short, in the *retention of my practical experiences a new form of 'protention'* is elaborated in which the lived meaning of the temporality of the instant *is determined within a project* that outlines the very form of being. In the experience of groping, the spatial form of the ambient world first appears in, to be sure, only an evanescent manner in the course of the movements that I accomplish even though these movements take on, in their real content, the aspect of real *work* by which I construct the surrounding space. But such a signification that is presented on the lived plane as only *passively preconstituted* is reinforced by the progressive accumulation of real experiences. At a given moment, the movements of displacement are inserted in a new form: they take on the meaning of displacement in a space that, at least in its general structure, has already been dominated. *The gaze of consciousness then returns to the meaning that has been preconstituting in retention and imposed as the actual reality of all past activity*. This the ego *once more achieves* in an explicit manner by consciously positing it as the very meaning of spatiality, obscurely intended in groping and now revealed in its true being as *geometrical structure.*[8]

Certainly, it will be said that a description of this kind concerns only the manner in which the intelligible is revealed to us (*pros emas*) and that no argument can be drawn from it against the anteriority of the intelligible in itself (*te phuse*). And, indeed, it seems that the movements that I gropingly have made were only possible by virtue of the very structure of space. To put it more generally, all truth *qua* truth presents itself not only as independent of the situation in which it arises but also as conditioning the very possibility of this situation. But such a *presentation* is only the result of the *self-forgetfulness* that belongs to the very essence of truth. In fact, the value of ideal forms is not separable from the movement in which they are constituted. The development of modern science has precisely taught us that no conceptual notion can be defined outside of the practical operations of verification in which it rightfully acquires its actual meaning. Thus, take the example of space: the flowering of non-Euclidean geometries and their application in physics demonstrate sufficiently that spatial structure is not an *in-itself* that is independent of *spatializing activities* of displacement and manipulation, taken in their ideal possibility as in geometry or in their application in actual practice as in physics. As soon as it is revealed that the measuring operations can no longer be performed in the customary manner at a certain scale of magnitudes, we realize that it is the very conception of space that the scientist must overthrow. The dialectic by which the sensible is itself transcended in order to be suppressed in the intelligible is valid only to the extent that it is accomplished in a real manner and is renewed constantly from an always new given. Thus, in the final analysis, it is always in *real sensible work* that the most elevated notions find the ultimate foundation of their intelligible meaning and their truth value. But it enters into their very meaning to forget forthwith the movement that engendered them. Their becoming is an *in-itself* or a *for-us*, not a *for consciousness* that lives in it: because *for consciousness* truth imposes itself from the outside as an *in-itself.*

> It is only . . . this origination of the new object – which offers itself to consciousness without consciousness knowing how it comes by it – that to us, who watch the process, is to be seen going on, so to say, behind its back. Thereby there enters into its process a moment of being *per se* or of being for us, which is not expressly presented to that consciousness which is in the grip of experience itself. The *content*, however, of what we see arising, exists for it, and we lay hold of and comprehend merely its formal character, i.e. its *bare* origination; *for it*, what has thus arisen has merely the character of object, while, *for us*, it appears at the same time as a process and coming into being.[9]

Notice that the result of the genesis is not limited to the logical formation as such. In addition to the positing of the forms of *syntax*, the act of judgment posits forms of the *kernel* as well. If I say that "*the table is round*," I have not simply stated a logical relation of subject and predicate: I have introduced new significations into the world of perception itself. The table is no longer merely this sensible unity given in intuition. It functions as subject *qua substantive*. Similarly, the round shape appears only as predicate because it has taken on the meaning of an *adjective*. Hence, the sensible object arises within a new horizon in which it is immediately available for predicative operations. Its constitution, henceforth, proceeds according to the form of a *necessary unity* that posits it as already determined in regard to the functions of judgment.[10] Its objectivity, henceforth, implies a recognition founded in the concept that regulates the synthesis of intuitions.

Thus a new world appears – a world of truth where the Kantian analyses of the notion of *experience* are verified. Within the horizon projected by the judgmental act, the objects are perceived as already determined in their properties and their spatio-temporal situation. We are dealing no longer with '*subjective-relative*' characteristics appearing in the *Heraclitean flux* of immediate perception but with '*objective*' determinations, whose objectivity is defined specifically in terms of their independence of the subject's situation. We shall say with good reason in regard to this new perception, in which the *event* is presented under the aspect of a *fact* that is definable in terms of concepts, that it conforms, necessarily, to the conditions of the judgment, since the judgment is the precise act that gives to the representation its relation to the *object* at this level. The primordial act of predication does not posit logical significations in a world apart from ideas; it makes them emerge within the very world of intuition that appears under a new form as the *new form of being*.

The ego is no longer this moving and motley unity of immediate sensible life: as the bearer of truth, it necessarily posits itself as self-identical, its identity being the very guarantee of the sense of objectivity. Under these conditions it will be legitimate to say that the world is *in me* in the act by which I constitute it. I am conscious of myself as positing the world in its true being through a primordial apperception – not, to be sure, in the sense that my representations are always true but in the sense that they are *mine*, as belonging to my self-identity. Kant explains in the *Deduction:*

> I do not mean that these representations *necessarily* belong *to one another* in the empirical intuition, but that they belong to one another *in virtue of the necessary unity* of apperception in the synthesis of intuitions, that is, according to principles of the objective determination of all representations insofar as knowledge can be acquired by means of these representations – principles which are all derived from the fundamental principles of the transcendental unity of apperception.[11]

Inasmuch as the world is *in me*, it is but a moment of myself, and I recognize myself *in it* in the primordial consciousness that I have of myself as constituting act. But since I recognize myself *in the world*, I grasp myself as one of its moments: within this horizon of all valid existence, the self affirms itself as an existent among others, since, henceforth, this will be the definition of the very notion of existence. I take over a place in the world that is *mine* in order to be fulfilled in my *truth*. The constituting is rediscovered immediately as constituted within the result of the constitution. The act of positing is immediately self-positing in the form of passivity: empirical consciousness presents itself as the auto-objectivation of the determining consciousness. Thus, it became impossible to constitute the intelligible on the foundation of the sensible, since the sensible itself appears only within the intellectual activity that has posited it. Starting with the world elaborated within the horizon of the judgment, Kantianism was working backwards and thus prohibited for itself all thematization of antepredicative content which, notwithstanding, founds the truth of predicative life. The intuitive given, persistently taken in its objectified form within the framework of scientific experience, was necessarily interpreted as 'psychological', and the constituting could be defined only by the exclusion of all material content. Hence, the system centered around the notion of object, without it being at all clear how an actually real object could appear. For the analysis of formal conditions of objectivity had evident bearing on the *concept* of the manifold: yet, it reached in no way the manifold that was *actually given* to our intuition.

It was not possible, however, to sidestep completely the problem of the *real possibility* of the empirical use of the categories. In the famous footnote to Paragraph 26 of the *Deduction*, mentioned earlier, Kant resolves this problem by seriously distorting the original definitions. He discovers, all of a sudden, that space and time, which at first appeared only as the *forms of intuition*, nevertheless as *formal intuitions* implied the unity of a synthesis in which "the understanding determines sensibility" even though "it precedes every concept" and has been, for this reason, attributed in the *Aesthetic* to the sensibility itself. It obviously concerned the expression of a felt and lived unity in sensible life – the latter being conceived not as 'a rhapsody of sensations' but as an actual *praxis*. The *form of intuition* which 'simply gives the manifold' is the manner in which consciousness finds itself affected – the form of its fundamental position experienced in the *certitude* of its real life. The becoming conscious of this form, as the passage from certainty to *truth*, posits the manifold in the unity of a *formal intuition*. The structure of being is no longer simply lived and *passively preconstituted* in sensible becoming where it is elaborated: it is *actively constituted* in the object as the *result* that emerges from this becoming. By becoming conscious of the *manner in which I feel myself living and acting*, I place myself in the *truth of my life as lived*, as the eternal subject of a world of necessary determinations. Within this *new horizon of being*, particular judgments intervene to explicate specific cases of actual determination. But the conceptual synthesis as such is possible solely on the foundation of a primordial act that has posited the general form of the world at the very level of intuition as the world of truth. The understanding is not the faculty of concepts inasmuch as they would impose themselves upon the sensibility from on high: it is but sensibility itself affirming itself in the meaning of its being as authentic consciousness of self. Thus the very movement of the *Deduction* tried to free itself from formalism. By separating the One from the manifold, every means of understanding how the unity is *the unity of this multiplicity* was taken away: it then remained to search within *this multiplicity* for the primordial upsurge of the One. The relation of the *form of intuition* to the *formal intuition* opens up new perspectives for the synthesis of apperception, which in its primordial act 'precedes every concept' and can be founded only on the very movement of the sensible as such. The world of objective determinations, such as Kant defined it, extends only to the *surface* of things. It is the existent just as it appears to us within the banality of daily life. But the act that *constitutes* it plunges its roots into the *preconstituted*. The existent takes its meaning from a horizon that refers to the preexistent in the *dimension of depth*.

It now remains to specify the constitution of this *sensible reason* to which, in the final analysis, the very meaning of all *logical reason* refers. We arrive here at the result of phenomenology in which the real content of the analyses of consciousness revealed the demand for a self-surpassing.

CHAPTER FOUR

THE RESULT OF PHENOMENOLOGY

21. THE GENESIS OF ANTEPREDICATIVE EXPERIENCE AND ITS REAL CONTENT

The constitution of antepredicative experience was the constant concern of Husserl from the time of the *Logical Investigations* to that of the *Crisis*. Two strata can be distinguished concerning this: the first, elaborated in a *Transcendental Aesthetic*, defines, in a broad sense, the 'primordial' world given to the individual Ego, *solus ipse*; the second, described in a *Transcendental Theory of the perception of the other* (*Einfühlung*) refers to the significations of intersubjectivity that are anterior to universality, properly so called.[1] The *Transcendental Aesthetic* itself consists of two parts: on the one hand, the constitution of lived experiences as 'immanent objects' in the immanent time of internal sense, which is the subject matter of the lectures of 1905, published by Heidegger, and of group *C* of the unpublished manuscripts; on the other hand, the constitution of the primordial external object has been developed in numerous texts that extend over the whole of Husserl's career and of which the most recent have been assembled in Group *D*. Group *E* contains the major unpublished manuscripts that have been devoted to the theory of *Einfühlung*.

Lived immanence that has been thematized by the transcendental reduction, in the classical sense, is still, as has been noted, a *constituted* that itself refers to the absoluteness of the Living Present (*Lebendige Gegenwart*). Thus we must proceed to a new 'reduction' by 'bracketing' the immanent unity of lived experiences, which makes them appear as singular durations remaining identical with themselves in the stream of consciousness. And so we come back to actually pure *impressions* in the *primordial flow of the absolute Present*: it is in this flow that there will appear the first intentionalities that allow for the transcendence of the pure '*given*'.

Each impression, indeed, taken in its immediate actuality, is maintained in a *retention* that is itself *present* but retains it, to be sure, merely as what *has just now been*. In the passive intentionality of retention, what was a present impression appears more and more as *past*. In this way the first dimension of the internal sense is constituted: the Living Present becomes *'object'* for itself by *retaining* itself. It *appears* to itself in retention where the pure impressional given is constituted in a *lived unity* of *sensation*, which presents itself as self-identical in that gliding by which it continually plunges into the immanent past. Among the sensations so constituted, we should give particular emphasis to *kinestheses*, which motivate an organization of the sensorial given in a *sensorial field* (*Sinnesfeld*) where the moments not yet present are already announced in advance: thus *'protention'* appears as itself present, but a case in which the Present 'goes beyond' itself by directing itself toward its *immanent future*. With the intentionality of protention the constitution of the internal sense is achieved: the domain of lived immanence is constituted therein as the first form of 'objectivity' – each experience appearing as an 'object in me' that endures and as identical to itself in immanent time.

The external sense is constituted upon the internal sense. As a result of the intervention of kinestheses, configurations are detached from the sensorial field as perspective unities where the first meaning of externality appears. The 'phantom'[2] is presented as an object that is *external to me* and is outlined within my sensory field according to a particular style of perspective, where it remains self-identical throughout its displacements and deformations. But it is merely a matter of a simple unity of appearances, which has not yet acquired the substantial and causal reality that is characteristic of the 'thing'. We do not yet have real succession in the time of things but simply an unfolding of perspectives within the immanent time of internal sense according to a certain unity of style which founds the perception of the object as 'phantom'. But spatio-temporality along with the 'world of phantoms' is constituted as the form of 'real' externality. Even though it is not yet a question of 'real' externality, properly so called, the changes there take on a new meaning: they are no longer simple changes of perspective but movements of 'phantoms'. In the interplay of these movements, in which the action of the kinestheses must intervene *de novo*, a certain *style* that defines the substantial and causal unity of the 'thing' is constituted. Understood correctly, there is no question here of the categories of the understanding but simply of a certain intuitive regularity that is founded on *habit*: nevertheless, it implies a well-determined general form in which *'sensible reality'* is constituted as such.

The world of 'things' flowing from the primordial constitution becomes in its turn the foundation for the experience of the other. Indeed, within the primordial horizon, my own body presents itself with a singular meaning: this 'thing' is also *my organic reality* whereby I function as an *Ego*. The meaning of 'living body', whose persistent and primordial experience I have *here*, is transposed by analogy to the material body that I see *over there* and which appears to me, consequently, as 'another' living body in which 'another self' functions: in fact, I have the experience of the *over there* seeing that I myself have been over there; and *the other* that I see is none other than *I myself over there*. The first social significations are constituted upon the foundation of the experience of 'the other'. It is within this horizon that we must finally place the genesis of ideal forms which arise in predication and which create the subject matter of the *Transcendental Analytic*.

Such is the plan of the constitutive investigations, whose sole value depends, evidently, on the prodigious subtlety of its detailed execution. This brief survey is, nevertheless, sufficient to provide us with an idea of the methodical precision of the Husserlian procedures and to reveal, at the same time, the profound contradiction that is responsible for their failure. It is not by chance that the published works remained in practice programmatic, since the work of realizing these works was never judged worthy of publication by its author. It is all too clear that the genesis of antepredicative experience, the masterpiece of the *Weltkonstitution*, was posited in reality on a ground incompatible with the philosophical framework on which it had been conceived, owing to the scrupulous exactitude of the description. It was, of course, a question of considering the *real content* of the development of the animal or even the infant.[3] In a more general way, the *life-world* (*Lebenswelt*) can only be this real world in which we live. Under these conditions there can be no doubt that the *Transcendental Ego*, in the phenomenological sense of the term, must not be identified, in fact, with real historical man. We know, moreover, that one of the essential points of the Husserlian doctrine is this – that to every proposition of transcendental phenomenology there corresponds a parallel proposition in psychology, the transition being possible in one sense or another by a simple 'change of sign'. From this point of view, the contribution of the phenomenological description in terms of concepts is invaluable. For until now these concepts were always lacking in experimental research for a *methodical and positive genesis of consciousness beginning with life*. The explication of the elementary structures on the plane of lived experience provides a certain number of necessary mediations for the passage of behavior to the level of consciousness. And if we admit, as it has been demonstrated, that the ideal forms of the world of the spirit find in the movement of sensible life *not only their historical condition for appearing but also the foundation of their truth-meaning*, we see the outline for the possibility of a scientific study of human existence. In this study, *explanation* in terms of real conditions does not suppress the *comprehension* of ideal values but, on the contrary, allows for their foundation and criticism in the very sense of their authentic pretensions. But an enterprise of this kind implied a radical transcendence of the properly phenomenological horizon of lived analyses: it is no longer a question of setting aside the world in order to return to consciousness but rather of understanding the real movement by which nature *becomes* human by *constituting itself* as spirit.

However, we should notice that there would not be a question there of some sort of an extra-philosophical utilization of phenomenological concepts, which would now put in dispute the very foundations of phenomenology. The absorption of the point of view of consciousness in natural reality is the sole conceivable way to escape from the phenomenological impasse. The fact is that the descriptions of antepredicative experience on which the whole edifice of the *Weltkonstitution* had to rest, have not given any really satisfactory result, in spite of the unique mastery of the author with respect to intentional analyses. The reason for this is that the sensible world cannot be defined with precision on the plane of pure lived experience as such; for it takes its whole meaning from the structures of behavior. There is no doubt that the directions of the 'sensorial field' correspond to the behavior of *locomotion*; the externality of the 'phantom' corresponds to that of *prehension*, and the substantiality of the primordial 'thing' corresponds to that of *manipulation*. Thus it is no wonder that here the phenomenological method encountered insurmountable difficulties: the description of antepredicative significations refers back to the conditions of material existence and necessarily places the subject within the framework of objective reality.

This will appear even more clearly, provided that we consider the syntheses that mediate predication within the life-world. The notion of 'explanation' as an antepredicative activity producing the relation of 'substrate' to its 'determination', corresponds to a definite phenomenological experience but remains intolerably obscure on the theoretical plane: how can the simple direction of one's attention extract a 'determination' from the undifferentiated unity of the 'thing' without an act of abstraction, which already presupposes the functions of the understanding? We do not see how a pure analysis of consciousness can escape contradictions of this kind, which originate precisely within the reflexive method itself. In fact, the movement of antepredicative 'explanation' defines reflection within consciousness as an instrumental function within the subject who makes use of a tool. Thus, the function of the stick is to lengthen one's arm. The 'explanation' of the determination is realized in behavior itself and endows with new significance the object that appears *in the passivity of retention* as the bearer of this determination. The phenomenon is absolutely general: the use of tools characterizes human existence as such and gives rise at each moment to new relations that are *passively preconstituted in the life-world* and which, subsequently, are affirmed actively in the acts of commandment and teaching on the symbolic level of predication. In other words, since *sensible life* for man does not consist in immediate interactions with the environment but is mediated by the *production* of the conditions of his existence, the passage from the sensible to the intelligible can be correctly described only by the analysis of the *technical* and *economic* forms of this production. This is, moreover, what Husserl obscurely presaged when, in a famous passage of *The Origin of Geometry*, he sought to found geometrical truth on human *praxis*:

> It is clear that the method of producing original idealities out of what is *prescientifically* given in the cultural world must have been written down and fixed in firm sentences *prior* to the existence of geometry; furthermore, the capacity for translating these sentences from vague linguistic understanding into the clarity of the reactivation of their self-evident meaning must have been, in its own way, handed down and ever capable of being handed down . . . . Only in this case could every geometer be capable of bringing to mediate self-evidence the meaning borne by every sentence, not merely as sedimented (logical) sentence-meaning but as its *actual* meaning, its *truth meaning* . . . . Further, it is clear that in the life of practical needs certain particularizations of shape stood out and that a technical *praxis* always (aimed at) the production of particular *preferred* shapes and the improvement of them according to certain directions of gradualness – among the lines, for example, straight lines are especially preferred, whereas totally or partially curved surfaces are undesirable for many kinds of practical interests. Thus the production of even surfaces and their perfection (polishing) always plays its role in *praxis*. So also in cases where just distribution is intended. Here the rough estimate of magnitudes is transformed into the measurement of magnitudes by counting the equal parts.[4]

Thus phenomenological explication was oriented toward the determination of the real conditions in which truth is engendered. But his subjectivist point of view in this area prevented Husserl from going beyond the level of commonsense remarks. What should have been done was to specify systematically that for each era *the real form of existence* inasmuch as it was experienced in a *certain intuition of the world* was expressed on the ideal plane by a *spiritual form*. In this way alone, the task of a total justification of being in the fullness of the meaning that it has for us, would have been accomplished: but such a description went beyond the limits of a description of lived experience.

It will be countered that the transition to real analysis would contradict the very notion of a '*transcendental constitution*'. But the contradiction is only that of phenomenology itself. The return to consciousness was motivated originally by the theoretical insufficiency of the positive sciences. It was a question of absorbing them in an *absolute knowledge* that would restore their authentic truth-meaning. Yet, Husserl's last analysis, in virtue of his genetic method itself, concludes in fact with a total skepticism:

> This universe of determinations in themselves whereby exact science grasps the universe of the existent, is but a garment of ideas thrown over the world of intuition and immediate experience: the life-world (*ein Kleid von Ideen, geworfen über die Welt unmittelbarer Anschauung und Erfahrung, über die Lebenswelt*) . . . . The return to antepredicative experience and the clear comprehension of its most profound and most primordial layer, has the meaning of a *justification of doxa* (*Rechtfertigung der Doxa*) which is the domain of completely primordial evidence not yet idealized in the 'exact' notions of mathematical physics. In this way it is also demonstrated that this domain of *doxa* does not imply evidence of a lesser rank than that of *episteme*, that of predicative knowledge and its results: it defines in a precise way the completely primordial domain, to which, by its very meaning, exact knowledge refers and which must be characterized as a simple method and not as a kind of knowledge which would mediate a thing in-itself.[5]

Hence, the relativity of knowledge must be understood in a sense which factually assimilates all kinds of truth:

> The trader in the market has his market-truth. In the relationship in which it stands, is his truth not a good one, and the best that a trader can use? It is a pseudo-truth, merely because the scientist, involved in a different relativity and judging with other aims and ideas, looks for other truth – with which a great many more things can be done, but not the one thing that has to be done in a market?[6]

Such a position implies the supreme consequence that the imagination not only has the same rights as the understanding but also must be preferred to it. This is precisely what Husserl asserts in a daring refutation of the Copernican system: the Earth is not a 'body' among other bodies; it is the absolute place of our primordial habitat, the immutable ground on which all meanings of motion are defined:

> There is only one humanity and one Earth – to which belong all fragments which are detached from it, or that will ever be detached from it. But if this is the case, have we the right to say with Galileo: "*par si muove*" and not, on the contrary, that it does not move? Not, to be sure, in the sense that it would be at rest in space while it could move therein, but as we have tried to show earlier, it is the first substrate (*die Arche*) which alone makes possible the meaning of all motion and all rest as a mode of motion. Its being at rest is not a mode of motion.[7]

Such an inversion of every sense of truth has been possible only through a fundamental error in the original project. It was an authentic question of rediscovering *the thing itself*, the true reality whose real meaning, abstract discourse had hidden from us. According to these conditions, the *subject* to which we returned could be none other than *real man* himself, who has to be awakened to the consciousness of his authentic being. But the *world of appearance* had monopolized (in ordinary language) the whole meaning of the notion of *reality*, and philosophical conversion was necessarily defined at the outset within the very horizon of daily existence. Since appearances imposed themselves there as the real world, their elimination was presented as a *bracketing of this world*; and the constituting movement which was thematized by the suspension of the *constituted being* (i.e., the *authentic reality* to which we returned) paradoxically took on the form of the *unreality of a pure consciousness*. But, correctly understood, that was only a *transcendental appearance* which reflected negatively on the prestige that still remained of the *world of appearance*. The phenomenological description of the constituting subjectivity, owing to its rigorous fidelity to lived experience, had to reveal its *actually real foundation*. The world that was set aside at the beginning in view of its *constituted* form is rediscovered as the *very reality* of the *constituting*:

> The Earth itself and we men, I with my body, belonging to my generation, my people, etc. . . . with all this historicity (*Geschichtlichkeit*) belong inseparably to the Ego and in a manner which is essentially unique and irreversible: all that is, refers to this historicity of the transcendental constitution (*diese Historizität tranzendentaler Konstitution*) as to the kernel which is proper to it and of which it is but the development – in other words all new possibility which is rediscovered for the world is connected with the meaning of being already acquired.[8] [i.e., the life-world as reality on this Earth is the "meaning of being already acquired," the "kernel" upon which ideal forms develop.]

But the discovery of the real content was achieved within the *horizon* of the phenomenological reduction. Since the existence of the world had been suspended because of the problematic of knowledge, there was a contradiction in presenting the real circumstances of life as it is lived as primordial moments of the *constituting* subjectivity. The whole meaning of constitution seems to fall into a complete relativism: since the transcendental foundation consists in this *terrestrial life* itself, the life of the trader will justify the market-truth, just as the life of the astronomer will justify astronomical-truth. Such is the ineluctable consequence of a philosophy of lived experience when developed in all of its rigor. The temptation would be great to return to formalism, which was protected, *a Limine*, from every adventure of this kind by defining the constituting to the exclusion of the concrete content. But such a *regression* would restore only the very difficulties that motivated the phenomenological thematization of actual consciousness: we would condemn ourselves to a new 'search for the concrete' which would only place us once again before the present impasse. On the other hand, this situation is intolerable and could not be prolonged. Thus, we must go ahead, by *accepting* with its authentic meaning the result that has just imposed itself on us. The absurdities that seem to flow from it are simply because of the persistence of the horizon in which the movement that elaborated it was brought about. The return of the constituted to the constituting in transcendental consciousness implied the exclusion of all objective reality. Since the truth of the world could not be founded on a moment of this same world, the position of the Ego in the world was interpreted as a pure contingency. And thus, the recognition of its real content which was imposed by the very faithfulness of the analysis, appeared, necessarily, as a scandalous transformation of the empirical conditions of existence into transcendental conditions of truth. But the scandal supposes that one has replaced these circumstances within the already abandoned framework of the world of the natural attitude. But, in fact, the 'natural reality' that is discovered in the depths of lived experience, *is no longer that which was presented to spontaneous consciousness before the reduction*.

The 'given' world of the natural attitude is a horizon of objects, each of which implies a meaning for being that places it within an already determinate 'region': a material thing, a living organism, or a human being. Between these regions no passage is conceivable, unless it is by virtue of being a pure negation. Thus, the matter worked on by man is no longer matter but a 'cultural object' – a tool, a book, or a painting – matter being itself but pure inert reality. Thus the domains are separated: the movement always already presupposes the meaning of the event to which it must give birth. Thus the relations between nature and man imply the already acquired presence of human signification, for man is explained only by man: *anthropos anthropon genna*. Such is the commonsense point of view that Aristotle's formula has established for the metaphysical tradition. The difficulties that he engenders are motives for a return to the constituting self. Naive consciousness, indeed, falls into this contradiction of taking significations for things in themselves; while the separation of kinds assures them a proper being, ideal and distinct from 'reality' as such, the latter finding itself reduced to an empty existence that derives its whole *sense* for being from an act that transcends it. Matter itself will have its meaning as 'physical' reality only through the categories that posit it as such. Hence, the authentic subject is no longer the self as a reality among other realities but the pure consciousness of the self whose act founds the object in its objective meaning. Thus the truth of the world refers to a transcendental self that is beyond the world. But a separation as abstract as this, presupposes that *being in-itself* is incapable in its own right of taking on a signification. Such a postulate will seem incontestable provided that one recalls that there is no knowledge other than the knowledge *of being* and the meaning that being has *for us* has for its sole *truth*-value the meaning that it has for and *in itself*. The separation of being and meaning holds, in fact, for the abstraction of different modes of existence – an abstraction which makes the *meaning of the being* of each domain appear to be independent of *being* itself. Posited in its abstract ideality, meaning refers immediately to a pure constituting subjectivity. But careful study of the latter reveals it to be a *movement* that envelopes the opposed kinds within a dialectical totality that is identical with respect to its actual content to the *historical movement* of reality itself. Thus, the latter is no longer a mosaic of separate regions, as it appears to be to naive consciousness. It now presents itself as the *concrete becoming* where being constitutes itself within the diversity of its abstract modes. Thus, there is no longer any valid reason to refuse to constituting subjectivity its predicates regarding reality. More precisely, it is *nature* itself *becoming-subject*. Materiality is no longer this particular region, which, in its abstract meaning, seemed to be incompatible with thought: it is the primordial and *ever present* form of the movement that elevates it to life and consciousness. We do not fall back into vulgar realism whose "specific limitation lay in its inability to comprehend the universe as a process – as matter developing in an historical process,"[9] On the contrary, we end with *dialectical materialism* as the *truth of transcendental idealism*. Since the naive attitude has been definitively suppressed by the reduction, the *practice* of the description of pure lived experience is necessarily *absorbed* within a dialectical materialism that *suppresses* it in its properly phenomenological sense in order to preserve it in its resultant form and *to elevate it to a superior level*. We maintain the authentic demands of the *Weltkonstitution* by getting rid of its idealistic illusions. It is, indeed, a matter of 'bracketing' the world of *constituted appearances*, which the *fetishism* of naive consciousness takes for realities in themselves, and of returning to true being through the *constituting subjectivity*. But the latter is not the *Heraclitean flux* of pure consciousness: it is the real movement by which nature *becomes conscious of itself* in biological evolution and human history. *Lived experience* is but an abstract moment of *real life*. It expresses the rhythm of becoming in which the forms are engendered. The merit of phenomenology was to have subjected it to a methodical description of a rare precision, and its conclusion that the sensible is the foundation of all truth. But the abstractness of its point of view prevented it from seeing the material content of this sensible life. It then remains but the pure 'sensible given', exclusive of every norm of truth. Hence, with the assertion of a radical contingency the whole edifice of the *Weltkonstitution* collapses. The total justification of the phenomenological undertaking, which would finally place philosophy 'on the sure road of science', ends, paradoxically, in a sensualistic relativism. With it, the great tradition of rationalistic idealism comes to an end: *realizing itself by suppressing itself*.

PART TWO

THE DIALECTIC OF REAL MOVEMENT

"The material world that is perceivable through the senses and to which we ourselves belong, is the only reality: and our consciousness and thinking, however suprasensuous they may seem, are only the product of a material bodily organ – the brain. Matter is not a product of mind, but mind itself is merely the highest product of matter."

(Engels, *Ludwig Feuerbach*)

INTRODUCTION TO PART TWO

1. CONSCIOUSNESS AND MATTER

"Now as to thought," says Leibniz in a famous passage in the Preface to the *New Essays*, "it is certain . . . that it could not be an intelligible modification of matter or one that could be comprised therein and explained; that is to say, that the feeling or thinking being is not a mechanical thing like a clock or a mill." And, indeed, if one were to enter such a machine, one would find but "sizes, figures, and motions"[1] but nothing that resembles a consciousness.

The formal evidence for such reasoning raises some suspicion in regard to its actual content. In fact, its whole force amounts to a simple *petitio principii*: having granted that it *goes without saying* that the concept of materiality is exhausted in that of a pure 'mechanical conjunction', the separation of thought follows with the necessity of a tautology. But the premise is arbitrary and based solely on a prejudice. And, in fact, the conception of the world as a gigantic machine arises only at the beginning of modern times, with the birth of capitalistic civilization, in which the whole of reality is reduced to its pure monetary equivalent, and is inserted as a simple object of measure and calculation in a universal mechanism. Thereby were severed the last ties which, through various modes of mystification, were still maintained between man and nature in the conceptions of the East and of Classical Greece. Even in Platonism, where the abstractive effort of Greek thought reached the notion of a purely supra-sensible reality, the Form of Beauty still bore witness to the ever living presence of a nature that remained itself by negating itself. The mechanical conception of the world gained credence with the first successes of modern physics, which defines material reality as an inert mass, motion itself being but a mode of inertia. Correlatively, thought was presented as pure interiority devoid of extension. The new science and philosophy developed from a presupposition of a separation that rendered unintelligible the existence of a world as *human world*. Nature lost its soul, and spirit lost its reality. Nevertheless, the unity reappeared in the form of a mystery. Whatever the intrinsic evidence of the *cogito*, it was impossible to suppress the testimony of practical life, in which there is constant affirmation of the experience which is paradoxical according to Cartesianism, of an extended soul and a thinking body. Descartes writes to Princess Elizabeth,

I consider that there are within us certain primitive notions which are like originals, on the model of which we form all of our knowledge: . . . for the body in particular we have but the notion of extension . . . for the soul alone we have but that of thought . . . finally for the soul and the body together we have but that of their union, which depends on the power that the soul has of moving the body, and the body of acting upon the soul.[2]

[But] to conceive of the union which exists between two things, is to conceive them as one thing . . . . Since Your Highness remarks that it is easier to attribute matter and extension to the soul than to attribute to it the capacity of moving a body and to be moved by it without possessing matter, I beg her to freely attribute this matter and extension to the soul.[3]

It would have been logical, under these conditions, to take the distinct notions of thought and extension as simple abstractions: they concern "the soul alone" and "the body in particular." But there remained the separation that was necessitated by the horizon wherein modern science developed. Scientific *objectivity* was established precisely through the radical elimination of all *subjective* signification. After Descartes, metaphysical rationalism gave itself the task of radicalizing the dualism and suppressing the mystery of the union in man by referring him to God, conceived as absolute substance or as the source of harmony. In the revolutionary period, philosophical reflection liberated itself from theology in order to absorb all singular value in the universality of an abstract objectivity that took on the double form of a *transcendental Ego* and a *pure materiality*. The opposition between mechanistic materialism and transcendental idealism covered over their necessary correlation *in abstracto*, where the alienation that permits scientific thought to establish itself on the plane of positive research was achieved. The development of technology was thereby assured. But the world became thereby unintelligible in its totality: how could the *object* take on a *meaning* for human life, and how could man himself emerge as an *objective reality*? The very definition of the object, as it imposed itself during the classical period of the rising bourgeoisie and as still dominates Western thought, posited as its principle the *absence of signification* as *the signification of reality*.

Nevertheless, a horizon such as this is exposed as incompatible with the *actual content* of positive investigations. Nature, which was presented as a pure mechanical development in space, was revealed as, in fact, an immense *becoming* in which matter is elevated to life – the evolution of life leading to humanity. The separation of kinds appears as a simple abstraction, every *form* emerging as a *historical formation* in a universal process. The opposition between man and nature no longer has a *raison d'être* – the discontinuity being but a *result* that does not exclude, but rather implies the continuity of becoming that has constituted it. In such a movement, each form is presented with its own originality that finds its precise foundation in the very process of its constitution. Thus, material phenomena, such as they developed after the cooling of the earth's crust, have resulted in a biological form in which the equilibrium of the physico-chemical relations defines a qualitatively new stage. The domain of life, once it has been constituted as a *historical result* of the anterior movement, becomes, in its turn, the theatre of a new evolution, in which the living forms become more and more differentiated by virtue of the very conditions of the process of cellular formation through the interchange of matter. But here again the actual reality of the scientific content suppresses the abstract and mechanistic postulates that seemed to condition it. For it is in no way a matter of denying the originality of specific forms, but rather of defining them using their authentic meaning by integrating them as moments in the total movement.

There is no longer any reason, under these conditions, for maintaining the restrictions and reservations by which bourgeois philosophy, from its inception, strove to limit the value of science. In the abstraction that marked the beginnings of the experimental method, the meaning of life as it is lived (i.e., as properly human life) could only be found in the pure *cogito*, where the putatively transcended religious effusions reappeared. Kant himself, by freeing speculation from metaphysical dogmatism, thought it necessary, nevertheless, to restore faith because of the exigencies of practical reason. The remnants of theological concepts marked the limits of classical bourgeois thought, which justified the new science only within the very framework of medieval traditions. The latter were interpreted, however, in the sense of going beyond and being freed from them. The restrictions that motivated the return to an absolute transcendence appear, from this point of view, as simple conventions that concealed a robust faith in the universal import of scientific explanation; and the traditional theses of freedom, the immortality of the soul, and the existence of God, did not in any way present the application of determinism to the phenomena of life as well as to those of the spirit. Kant states, in a famous passage in the *Antinomies*, " . . . and if we could exhaustively investigate all the appearances of men's wills, there would not be found a single human action which we could not predict with certainty, and recognize as proceeding necessarily from its antecedent conditions."[4]

Thus, classical bourgeois tradition, according to its authentic intentions, aimed, indeed, at the constitution of a total science, and it would seem logical that the development of positive investigations that opened up concrete perspectives for a fully rational comprehension of the world was enough to eliminate the theological restrictions that could have been imposed at the beginning; *alienation* is resolved in the very *result* of scientific work – i.e., man recognizes himself in nature of which he appears as its supreme moment.

Nevertheless, we know that reality was exactly the opposite. As soon as the development of capitalistic structures had exposed the menace of a proletarian revolution, bourgeois thought entered the path of a systematic reaction. If, on the one hand, Auguste Comte still thought that he had to argue from a position of so-called '*positivism*' in order to reduce the *positive method* to a rhapsody of detailed studies deprived of signification, Schopenhauer, Kierkegaard, Nietzsche, and Bergson (on the other hand) engaged in an unremitting struggle for the return to mysticism, to pure interior feeling, or to unreflective action. It was the singular merit of Husserl to have turned away from these *regressive forms of behavior* in the name of the traditional demands for rigorous method and strict rationality. From this point of view, he represented the ultimate resurgence of the great classical tradition, and the scrupulous exactitude that he observed in intentional analysis brought him to the threshold of dialectical materialism. This aspect of phenomenology remained, it is true, totally misunderstood – the public wanting only to retain Husserl's hostility to naturalism with some intuitionalistic formulas which permitted, apparently, the rejection of scientific explanations in favor of pure significations as such. The prospect, to the advantage of the notion of essence, of being able, henceforth, to define the *true being of consciousness* by using the idea that consciousness itself has of it, and of resolutely taking the *intention for the fact*, aroused a prodigious enthusiasm that had no equal other than the absence of real curiosity about Husserl's own texts. The positive result of the Husserlian descriptions, which implied, in fact, a merciless critique of pretensions, experienced ideality in regard to the material content of sensible experience, was systematically neglected, and the phenomenological school, along with existentialistic interpretations, fell into a total irrationalism.

Such a decline implied an obstinacy bent on rejecting the new horizon elaborated by the dialectic of scientific thought, in which the objections raised about the 'problems of quality' are found to be refuted by the constitution of all diversity within unity itself, as the *passage* from quantity to quality. This is what we can verify on the level of properly material phenomena. The most important investigations of contemporary physics rested on the equivalence of mass and energy, and we might see here, considering this matter only superficially, an ultimate 'reduction' to identity. Yet, it is, in fact, a matter of constituting the multiplicity within unity, since we conclude by explaining the specificity of chemical substances by means of pure relations of energy. Undoubtedly, sensible qualities that are properly experienced will be understandable only on the human level. Here, it is still merely a question of the originality proper to the chemical domain, which is founded on the relative permanence of the distribution of elementary particles within the molecule – these representing but a condensed form of energy. Autonomy is not less real even though it is always provisional, as is attested to by the phenomena of materialization and dematerialization. Chemical properties result from relatively stable formations that are constituted in the development of phenomena of energy. Thus, the passage from the physical to the chemical implies at the same time identity and difference, as identity of the identity and non-identity.

The study of the phenomena of life, in its turn, demonstrates the dialectical character of scientific thought. The biological is explained exclusively and in all rationality in terms of the physico-chemical. Moreover, it could not be a matter of a pure and simple assimilation, since the living forms, as *particular forms* in the development of physico-chemical phenomena, imply an incontestable originality because of the relative stability of the process wherein they are constituted. Such a stability does not come from a transcendental intervention or some 'vital force'; it necessarily results from the very movement of properly material actions and reactions, inasmuch as they attain a certain form of equilibrium.[5] Thus, what is maintained is but provisional, the living form evolving from its physico-chemical infra-structure. The fact of heredity does not exclude, but rather implies variation – the germinating cells being but an abstract moment in the organic totality, and being modified necessarily according to the general movement of the adaptations of the living organism, at least to the extent that these influence the chemical composition of substances in circulation.[6] These transformations, nevertheless, are presented in regard to their *result*, in an original form, as the *evolution of species*. Inasmuch as life *transcends* matter, it *suppresses* matter by *preserving* it – the negation resulting necessarily from the preservation, and the preservation being implied in the negation. It is not at all a matter of an ideal dialectic such as we find in Hegel, but of the very movement of the scientific explanation, inasmuch as it gives an account of biological phenomena, precisely in their properly biological characteristics, in terms of physico-chemical phenomena. Such a dialectic does not superimpose a 'philosophy of nature' upon positive science; it defines only the *truth* of the scientific concept to the extent that it reproduces in consciousness the *real* process by which life is constituted in the general movement of material structures. Thus, on the strength of its content, a new way of thinking is imposed in which the objections that bring grief to science by suppressing the specificity of the different domains of being disappears. Such arguments are possible only in terms of the abstract conception of identity, where all difference is transformed into a metaphysical absolute.

The law of identity [says Engels in *Dialectics of Nature*], is the fundamental principle of the antiquated conception of the world: $a = a$. Everything is identical to itself. Everything was deemed permanent: the solar system, stars, organisms. This principle has been refuted by natural science bit by bit in each separate case, but in theory it still prevails and is still put forward by the supporters of the old in opposition to the new: a thing cannot simultaneously be itself and something else. And yet the fact that true, concrete identity includes difference and change has recently been shown in detail by natural science.*[7]

It is within this new horizon that it is appropriate to take up again the old problem of the relations of matter to consciousness. Leibniz's argument now appears to be meaningless. He presupposes that matter is incapable of evolving: yet the reality of evolution is no longer in doubt, and matter *becomes* thereby something other than itself – i.e., precisely life and consciousness. Naturally such a formula does not resolve the problem in any way: but at least it poses it in intelligible terms that open up a path for positive investigation. It is a matter of searching methodically; and by this real process, matter that has become living is elevated to spirit. Reality *qua* dialectic is movement of such a type that, in each mode of being, the changes of causal order that are determined by the very structure of this mode *necessarily* conclude, by the very development of their content, in the constitution of a new mode that *absorbs* the first and maintains it as suppressed, preserved, transcended. It is the very continuity of becoming that determines the discontinuity of forms, since *mechanical* movement, developing within the causal framework, necessarily overflows this framework and is constituted in a *dialectical* movement as the passage to a more elevated type. Such a passage, in which the old form, negated in its external appearances, is constituted in its *true being*, defines the authentic meaning of necessity as the *realization of freedom*.

It would seem, then, that all that remains to be done is to enter into the details of the content. Yet a difficulty remains that can still hold us back.

Though one sees fairly easily how life can result from an equilibrium of material relations (more precisely, from an equilibrium of autocatalytic phenomena), the domain of the spirit is presented with an altogether different specificity. The living organism is still but an *object* among other objects, and none of the great thinkers of the classical period, from Descartes to Kant, has doubted the possibility of treating it as such. Yet, consciousness as consciousness *of the object* implies, it would seem, a 'transcendence' that elevates it above natural reality and absorbs it in the pure interiority of the self, immediately given in the *consciousness of self*. There would appear, consequently, a radical heterogeneity and not difference within identity. The subject is subject only to the extent that it is not an object. Yet, as far as we are concerned, it is a matter, of course, of describing this genesis *precisely qua subject*. The project, apparently, runs up against a contradiction that has no equivalent in the dialectic of nature. There could be no question, however, of realizing it within a metaphysical absolute and building the *absence of a solution* into a *philosophical solution*. The interiority of consciousness and its ideal relation to the object refer back to the structure of the *real subject*, namely the living body as the center of movements. It is not at all a matter of an arbitrary 'reduction' to the point of view of the object: the act of consciousness in its lived meaning is defined exhaustively by the dialectic of behavior. Here again, philosophical dilemmas are resolved within the very movement of positive explanation.

Idealistic analysis discovered long ago that all consciousness implies an activity on the part of the subject, but it used this fact only as a weapon against materialism – paradoxically transforming the actuality of human life into a simple negation of its reality. For its part, the objective method did not go beyond the mechanistic consideration of spatial models. The apparent opposition between subjective idealism and abstract materialism concealed their deep agreement in an equal refusal to thematize the concrete human activity as the ongoing subject of objective reality. Such a dialectic is, nevertheless, visible in any phenomenon of consciousness whatsoever that is taken in its *actual content*, inasmuch as it refers irreducibly to real movements within the living organism. When I see this tree, I feel, more or less confusedly, a collection of reactions that sketch out a horizon of *practical* possibility being outlined within myself – e.g., the possibility of getting near to, moving away from, circling, climbing, cutting, or uprooting it, etc. The lived meaning of the object (i.e., its being-for-me) is defined by these very possibilities that are lived and felt in these modes of behavior that have been outlined and that are immediately *repressed* or *inhibited* by the objective data – the real act being reduced here to a simple oculo-motor adaptation. Consciousness, as consciousness *of the object*, is precisely nothing but the very movement of these repressed outlines. In this repression, the subject (i.e., the living organism) preserves them *in themselves*, and it is this very preservation that constitutes the *consciousness of self*. So, it is truly in its lived meaning, and not simply from an 'external' point of view, that consciousness is defined in terms of the dialectical of behavior. The reactions awakened by the stimulus and arrested by the real act before they could have reached the phase of completion are integrated within the total behavior as *moments* that are *suppressed, preserved, transcended*. In this movement, the living organism has become subject. Life consists originally in a *real* absorption of environmental matters. But from this time on, the development of this process leads the living being to extend to its surroundings a collection of reactions that have been outlined in advance. In this anticipated absorption of the object, the subject possesses it in himself; in other words, he is *conscious* of it.

Such is the reason why all consciousness is identically consciousness *of the object* and consciousness *of self*. I am necessarily conscious within myself of the meaning of the object *for me*, since this meaning, because it is actually lived, is nothing but the very movement of forms of behavior that have been repressed *in me*. But we see, thereby, that the subject is not immediately conscious of its real behavior, which, nevertheless, defines it in its true being. The intentional meaning includes solely the acts that are outlined and their silhouetted form as lived pretension. But the reality of the act wherein the movement ends and wherein the outlines with their pretensions are absorbed is not itself lived in its original manifestation.

This is the truth that the critical school presented obscurely when it denied to the subject the possibility of knowing himself, for all knowledge implies an *object*, and the subject *qua* subject could not be given to himself as *object*. This observation is perfectly correct if one subscribes to the reflexive method: for the actual reality of the self consists in the real content of its behavior, which escapes, to be sure, the consciousness that it has of itself. One must simply conclude that the plane of consciousness has to be transcended and the self defined in terms of the object, since positive analysis alone realizes the actual conditions of the old philosophical project of a knowledge of the self. The objection will be raised, it is true, that a definition of this type would not allow the subject to *recognize* himself, and that recognition is implied in the very essence of subjectivity. We shall reply that contradictions result simply from the fact that one persists in taking concepts from a metaphysical point of view that is outside of the movement of their content. For the becoming of the real object (viz, the living organism in its actual behavior) makes him, accordingly, *pass on* to consciousness. The exterior becomes the interior in a movement in which the consciousness of self is constituted as the consciousness that reality has of itself in its truth-meaning.

Here again it is not a question of a dialectic of concepts. The becoming subject from the object, as a production of difference within identity, supposes nothing more than the very activity of behavior. Every act, once accomplished, leaves its trace in the form of a disposition of the living organism to *repeat* it; not, of course, in its particular details, but in its general structure. The development of such a disposition ends in a new scheme of behavior in which the old form is reproduced in a collection of outlines, repressed and lived as an intentional movement that defines the object with its true meaning. If I grope about in a dark room, the object has at first, for me, only the indeterminate meaning implied in the groping. But the movements that I accomplish necessarily take on, by their very actualization and outside of my consciousness, a certain determinate structure through the objective relation of my body to the ambient bodies. My real acts are movements of a certain form, even though I am not conscious of them as movements of that form, but only as groping movements. Thus, indeed, it is true that the real act is not originally experienced in its actual reality, but it will pass straightaway into consciousness through the dialectic of behavior. For, to the extent that I accomplish these movements, there is constituted in my organism a disposition to *reproduce* them in the form in which they were accomplished, and this disposition corresponds to the structure of the environment. In this way, a new scheme is elaborated, defined in terms of a collection of anticipated outlines, which the real act represses and refers back to experience as a horizon of practical possibilities that give the object its sense of being as the space of a certain structure. In this experience the activity of the subject is no longer lived as groping movements, but as a movement *directed* towards an objectively determinate place. The subject thereby *becomes conscious* of himself just as he *already* really *was* at the beginning of the experience and just as he *recognizes* himself in his *true being*. Such a truth is evidently effective only at the *instant of the passage*. Since the lived act of becoming conscious is itself, by its actual accomplishment, only the result of the *repression* or *inhibition* of the old by the new form, this new form once again escapes to lived experience and sets in motion new development.

Thus, the consciousness of self by its actually lived meaning is truly but a particular moment in the becoming of the real subject. His *authenticity* refers to those privileged moments in which he coincides with his objective being. But the lived experience of this self-coincidence is evanescent, and the concept of it can be determined only on the plane of the object – the movement of the content serving to make it pass to subjective life. Only dialectical materialism enables the subject to understand himself and to dominate himself within the actual reality of his being as it first escapes his consciousness, but progressively imposes itself on him as his *truth-meaning*.

CHAPTER ONE

THE DIALECTIC OF ANIMAL BEHAVIOR AS THE BECOMING OF SENSE CERTAINTY

2. PHENOMENOLOGICAL GIVENS AND REAL GIVENS

In *The Phenomenology of Mind*, Hegel began his study of the genesis of consciousness with the immediate certainty of the particular object given as such to intuition. It is clear that a 'given' of this kind is 'immediate' only in a strictly relative manner: the 'this' intended in sensible certainty is already an *object*, and as such implies a form of intentionality that has been able to appear only after a long process of constitution. Conceptual dialectic was incapable of clarifying, on the properly sensible plane, the lived moment that Hegel had recognized, however confusedly, in the *Lectures* at Jena. Here is the place that the superiority of the intuitive method, as it was practiced by Husserl, was revealed. For the first time, the pure description of lived experience made possible the explanation of the significations that are constitutive of the sensible object as such. The 'thing' given in antepredicative perception already implies the moments of quality (i.e., spatiotemporal and substantial permanence): hence it can be understood only in terms of a *genesis* that goes through the stages of *impression, sensation, sensorial field*, and *object-phantom*. Correlatively, we have, on the 'noetic' plane and outside of the pure present, the 'immanent' intentionalities of *retention* and *protention*, and the first 'transcendent' intentionality as the *perception of the phantom*.

As remarkable as these results are, the reading of Husserl's text, nevertheless, leaves a profoundly deceptive impression, and a historical exposition of it would be disastrous. The apparent meticulousness in regard to detail only serves to underline the complete absence of motivations and actual determinations, and if phenomenology here reaches the ultimate limits of lived analysis, these very limits impose the necessity of going beyond them. But if we return to the positive results of the experimental investigations, we notice a striking convergence and one that, at the same time, enables us to make up for the insufficiencies of the Husserlian description: this description thereby loses its idealistic form in order to be absorbed within dialectical materialism as an abstract moment. Thus, if we take the fundamental phenomenological distinction between the 'phantom' and the 'thing', – the first being defined as a simple sensible exteriority, the second implying a sense of permanence that enables it to be grasped as existing even outside of the actual field of perception – we see that it obviously concerns all the evidence in regard to the stages, described by Piaget, of the sensori-motor development of the infant between five and nine months. Indeed, at the age of five months, the act of *prehension* appears, which reveals, for the first time, the perception of an object as external. This exteriority, however, does not yet imply any substantial permanence, since the infant is incapable of again picking up a rattle that has just been placed behind his pillow – this means that the object is, for him, only a 'phantom', which exists to the extent that it is actually present. The problem is resolved at the age of nine months, the age at which the perceived object has taken on the meaning of a permanent 'thing' for which the child does not hesitate to search behind the screen. On the other hand, if we refer to the evolution of animal behavior, we can notice the appearance of the structure of prehension, or *apprehension*, in the act of entrapment among fish. The behavioral acts of *detour* and of *manipulation* that correspond to the perception of the 'thing' are present in mammals. Thus we can consider the following sequence to have been established:

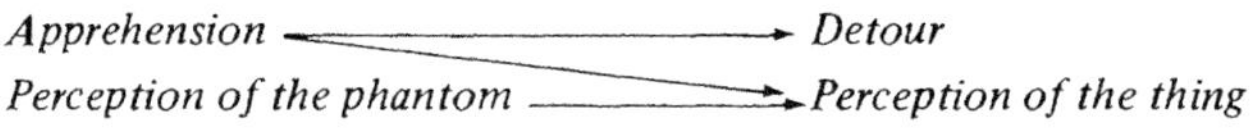

where one readily sees that the structure of real behavior at the preceding stage defines the intentional content of the lived act at the stage that follows. Indeed, to perceive an object as a permanent 'thing' is to *apprehend* it in a movement that has been outlined and immediately repressed and that is lived as an ideal movement by which I *keep* it in my consciousness in such a way that it preserves for me its sense of existence when it passes behind the screen.

Yet, the time-lag that takes place from the real act to the lived meaning of it can be verified in all its generality by a methodical analysis. Thus, the sense of exteriority that defines the perception of the 'phantom' is constituted, manifestly, in a movement of *locomotion* that is outlined and immediately repressed, by which I ideally *reach* the external object. Yet, the form of locomotion characterizes precisely the real behavior of the stage that immediately precedes the stage of apprehension: it appears in the animal series with worms, and in the development of the child at the beginning of the second month, when the infant begins to put his thumb in his mouth and directs his eyes toward the light. On the plane of consciousness, the act of locomotion corresponds, evidently, to lived orientations in the *sensorial field*. Yet these orientations appear precisely within the phenomenological series, immediately before the perception of the phantom. We thus have a new sequence:

Locomotion → *Apprehension*
Sensorial field → *Perception of the phantom*

If we go back even further, we have the stage of *reflexive displacement*, just as is present in phylogenesis in the coelenterates, and in ontogenesis in the new-born, and which corresponds, rather obviously, to pure *sensation* as such. Yet, we readily see that the intentional content of the orientations of the sensorial field can be defined in no other way than as a group of displacements outlined and immediately repressed:

Displacement → *Locomotion*
Sensation → *Sensorial field*

A more delicate problem arises for sensation. It is indeed preceded, in the phenomenological series, by the *sensorial impression* that corresponds in the real series to the elementary form of *contraction* in the way that it is observed in Porifera. Yet, the passage from impression to sensation does not, strictly speaking, have any difference of content: for sensation is precisely nothing other than an impression that lasts. The difference, then, applies only to the stabilization of impressions in the neuro-sensory cell. In both cases, it is an outlined and repressed movement of *attraction* or *repulsion* that send us back to the primordial structure of behavior in the Protista.

Thus, on the condition of taking into account a systematic time-lag, it is possible to make the phenomenological series and the real series coincide in a strict manner, since the intentional forms of each stage are identified with the real forms of the preceding stage. The lived meaning of consciousness comprises nothing more than the abstract movement of by-passed structures in the evolution of behavior: in other words, consciousness is the movement of *deferred behavior*, which sends us back to *deferred circuits* in the course of the nervous influx. The structure of the nervous system that results from the whole evolution does, indeed, reproduce the different stages that we have just outlined. The hierarchy of centers in the higher animals (spinal cord, cerebral trunk, and cerebellum, *corpus striatum*, cortex) does, indeed, correspond to the genetic series of behavior: reflex displacement, locomotion or orientation, affective automatism (apprehension), voluntary activity (detour and manipulation). But we know that, at each level, the inferior center is inhibited by the superior center, and this fact corroborates the definition of consciousness as an outlined and repressed movement. Thus, subjectivity is but the formal aspect of the real dialectical process, in which each new structure represses the one that precedes it and absorbs it in a lived interiority. Thus, there is nothing further to be done than to take up again the whole development: the real genesis of consciousness in which its material nature is demonstrated will enable us to explain, in a decisive manner, the authentic foundation of its *truth*-value.

3. THE MOVEMENT OF THE INTERNAL SENSE

The first forms of life, such as arise in the evolution of matter, obviously do not yet imply any consciousness. The behavior that appears in unicellular organisms or Protista directly results from the interplay of physiological processes and is defined, simply, as attraction and repulsion. The general concept of attraction and repulsion, owing to its form, does belong to behavior, since it implies the consideration of the living cell as a whole, but we cannot yet find therein any actual content that goes beyond the biological notion of *irritability*. But the constitution of the first cellular agglomerations sets in motion a new development: the composition of the elementary reactions of attraction and repulsion produces whole movements in which the first are absorbed. Subjected to the movement of waves, the colony that is fastened to the rock is constituted in a form of equilibrium whose rupture, in case of abnormal circumstances (such as when the disturbance of the waves stops) produces the movement of *contraction*. This contraction defines the *muscular function*, which appears for the first time with the Porifera. Sponges are agglomerations of cells that are still relatively independent, forming a kind of sack fastened to the base of the rock and opening up, at the opposite extremity, at the osculum. The whole is capable of contractions, which sometimes succeed in obstructing the interior cavity (Perrier). But the reaction is produced more specifically on the level of the osculum, in which a first outline of muscular fibers appears. In such a movement, the original irritability of the elementary cells takes the form of a process of *transmission* in which the *nervous function* is defined. Even sponges do not possess any nerve tissue, a phenomenon of neuroid transmission is revealed when the closing of the osculum is provoked by a wound several centimeters away. The nerve function appears, thus, as the *secondary form of behavior*: the original reactions of attraction and repulsion that are proper to epithelial cells are inhibited by the contraction of the osculum and thus are found reduced to the movement of transmission. Thus, the colony of Protista has evolved into an organic unity that reacts to external excitations by means of its own behavior. The sensory-impression, as the primordial form of lived experience, does not imply any transcendental mystery: it is only the irritability of the cellular element absorbed and repressed in the reaction of the whole organism.

The impression that is experienced in an instant as pure shock does not as yet imply any duration, properly speaking. At this original stage, the subject is but this pure, evanescent, and living present. Let us note, however, that even at this level, consciousness is defined with regard to its actually real content as a mode of activity. The impression is not an external 'given', but a repressed reaction. Neural transmission is not a passive transfer of excitation, but a moment proper to the living cell, a movement that results from its irritability. The impression, as a repressed movement of attraction or repulsion, bears a *meaning* that is defined precisely in terms of this very attraction or repulsion. But this meaning is lived only at the present instant. It is not actually possessed as *sensation*. The passage to sensation is effected by the exercise of muscular movement, in which the neural transmission develops. Since behavior is organized around the digestive cavity, where the activity of the living organism is concentrated, the ways of transmission stabilize themselves in particular elements where the nerve tissue is differentiated. In the neuro-sensory cell, the elementary reactions of attraction or repulsion that arise at the epithelial level take on a specific form and so are absorbed within the total movement, no longer as simple, instantaneous impressions, but as *unities of meaning*. While excitation is going on and while muscular reaction is taking place, nervous pulsations accumulate: these are repressed and interiorized in the form of a series of impressions that are maintained with their *still present* meaning of attraction or repulsion. In this retention, the attractive or repulsive *sensation* is constituted as a lived experience that endures, is self-identical, and progressively penetrates into the immanent past. At this stage, the form of behavior is still defined, in terms of the internal unity of the organism, as *reflex* activity. But the actual development of reflex immediately leads the subject to transcend itself in an objective manner in the act of *displacement*. The changes of position that are observed in Protista were but mechanical consequences of elementary reactions of attraction and repulsion; they did not have the form of displacement. This form is defined in terms of a more elevated dialectical structure, in which attraction and repulsion are repressed as soon as they are outlined, and maintained at the nascent stage as *suppressed, preserved, transcended*. This is the movement that is realized in the coelenterate where, with the nervous system, the first reflex acts appear: ingestion of food by the coordinated movements of the tentacles and the mouth, displacement by reptation (polyps) or jerky leaps (medusa). One also observes the first acquisitions of rhythm in relation to the periodicity of external phenomena, and these reveal the presence of an internal time. Thus, sea anemones that have clung to elevated rocks close shut again shortly before the low tide and avoid dessication by retaining water in their cavity (Pierón). Yet, to be sure, this is not the place to speak of a 'conditioned reflex', since conditioning in the strict sense implies a much more complex organization and one that will have been achieved only with worms. We do not have here the constitution of a new stimulus, but simply the formation of an internal rhythm that is based on the *retention* of the immanent past.

With the sponges and the coelenterates to which the ancients gave the evocative name of *zoophytes* (i.e., 'animal plants'), vegetative life has been elevated to *animality*, as sensorial consciousness and life of relation. But lived experience is still presented merely as pure *given* sensation. Even though the act of displacement implies a 'going beyond' the given as such, this 'going beyond' exists only for the objective reality of behavior. It does not reflect upon itself in consciousness. The real movements of the subject are objectively oriented by sensation that is present in it, but this orientation is not lived as such, and the subject is not conscious of the direction in which it moves. But the actual content of its movements develops within a new dialectic as the dialectic of experience: the subject *experiences* the environment in which it is displaced. The real structure of its displacement, which is first of all determined solely by the material relations of the organism with the environment and thus necessarily escapes consciousness, is *reproduced* in the organism on the level of nervous commands and in the form of a disposition for being displaced in a precise and certain manner to which it has become accustomed.[1] In other words, the displacements are progressively associated with the sensations corresponding to external excitations – notably luminous excitations, whose action already provokes in the medusa the formation of ocular spots. The movement is constituted, then, into a new formation – that of *locomotion*, which we can define as a displacement which is actualized among several others that have already been outlined, and arrests them by means of its very achievement. The outlines thus repressed are lived through in the form of a '*protention*', in which the subject is conscious of anticipating its immanent becoming. Hence, sensation no longer appears as a simple unity of duration that is given in retention; it is inserted within the *horizon* that has been drawn by the intentionality of protention, and is presented as a moment of the *sensorial field* in which the collection of repressed and lived displacements is manifested as possible in practice. Thus, the external environment is interiorized in a lived environment in which the subject *directs* itself. This is the movement that is realized with the worms, where the development of locomotion is translated, in the somatic plane, by the opposition of the anterior and posterior and bilateral symmetry. The nervous system, which controlled the displacements of the coelenterates, is maintained, still, under the epithelial layer, but simply as a *transcended formation*, its influx being absorbed within the central nervous system (i.e., longitudinal grooves and cerebroganglia), which insure the more complex movements of locomotion. Indeed, this transcended formation is presented as the *result* of the experience of displacement that has been summarized in the passage from the influx of the nervous system to the central system. On the anterodorsal face, now more exposed to excitations, there develop sensory organs (i.e., eyes and statocysts), whose impressions are projected on the cerebroganglia, where the direction of movement is concentrated. Thus, the sensory givens are no longer simply felt as attractive or repulsive; they are absorbed within the horizon of possible displacements and are constituted in a *sensorial field*, of which each moment implies a lived value for locomotion. Such a value defines the meaning of *need* and *tendency*: thus, the earthworm seeks damp and dark places. At the preceding level, affectivity consisted merely in sensation itself, whose sense of attraction or repulsion was experienced immediately as pleasure or pain. With the intentionality of protention, in which a sensorial field is drawn out, pleasure and pain are themselves *transcended*, in order to be constituted as field values: light appears to the worm as a negative value, and dampness as a privileged direction in the environment in which it moves. It is also this lived transcendence that founds the possibility of conditioning in the strict sense. The general fact of learning, as a disposition of the living organism to repeat its anterior movements, is an essential property that can be verified at the level of the Protista. But as such it does not have meaning for the subject, though it founds the precise dialectic of behavior in which subjective life is engendered. Conditioning, properly speaking, as Pavlov has shown, supposes a system of already constituted reflexes. Since the reflex act already bears an objective orientation, the acquisition of the conditioned reflex in individual experience is realized on the foundation of a horizon of possible orientations. In short, it is not a question of a simple mechanical association of isolated sensations, but of the formation, in the sensorial field, of a direction that presents a particular signification for the subject. Thus, when an earthworm learns to avoid a particular passage in a maze, where it has hit against an electric grating a certain number of times, we cannot speak of a simple repetition of former behavior: the worm does not reproduce its original reaction to the electric shock; it *avoids* a certain direction that appears to it as a negative value in the *field*.

We should note that the sensorial field has not made us depart from the domain of lived immanence. It is simply a matter of a pure internal environment where the subject is conscious of the direction in which it moves, but not of the *object* towards which it moves. The meaning of the object as external to consciousness implies the development of the actually real content of locomotion to the extent that it has not only a determinate direction, but also a terminal point; and to the extent that the subject *reaches* objects, there is constituted in its organism a disposition to reproduce its behavior in the form of movements of locomotion that are outlined as being achieved, from now on, in regard to tactile and gustatory sensations. In this new schema, the real act takes the form of *apprehension* that is defined as a locomotion being constituted on a first locomotion outlined as already achieved. Transcended by the act of apprehension, the movement of locomotion that is outlined is repressed and returned to lived experience in the form of an intentional movement that has reached, at that very time, the place towards which it is going. In such an anticipated achievement, the subject is no longer conscious simply of the direction of its movements as immanent future, but also of their terminal point as an *external position*, where the sensorial givens take the form of *external object*. To be sure, such an objectivity does not as yet have anything of the characteristic substantial permanence of the 'thing'; it is a question solely of a sensible exteriority – of a 'phantom' whose entire meaning consists in being attained from now on in a repressed outline of locomotion. Nevertheless, its existence is imposed within the limits of the movement that constitutes it. In fact, apprehension that is realized from the original attractions and repulsions comprises the forms of approach and flight, as well as that of attack, which has its origin in a combination of attraction and repulsion. In such a way are defined the behaviors of joy, fear, and anger, where the meaning of sensation is no longer simply lived in the interiority of the internal sense, but appears in the phantom that is presented as an attractive, repulsive, or irritating object. The emotion is not the simple, purely subjective, movement of need; it implies the meaning of the object as *moving object*. Nevertheless, the relation does not go beyond the limits of the present that is lived through. We do not yet have the stability of feeling in which the subject transcends its actual perception in order to relate it to a permanent reality. The whole being of the phantom is exhausted at each instant in its immediately given aspect, and emotional life consists merely in a scattered succession of immediate responses as *automatic activity*. Such is the dialectic that is operative in the passage to vertebrates. The behavior of apprehension appears in a primitive form in fish that hurl themselves upon and seize their prey. The perception of a distant object that develops with the organs of sight, hearing, and smell,[2] implies that the movement of simple locomotion is absorbed, repressed, and experienced as an intentional movement in which distance is already overcome and the object reached. One also notices the appearance of emotional life with the reactions of anger and fear in attack and flight, and the apprehension of food defining the primordial form of joy. Finally, the sense of exteriority enables the constitution, in its most primitive form, of the relation with the 'other' in group behavior. To be sure, it is only a matter of a simple drawing closer of individuals that move about together; the 'companion' is but a phantom whose evanescent existence is placed outside of all lasting feeling as of all collaborative activity.

With the going beyond and repression of locomotion through apprehension in fish, the nervous system acquires the general structure that it will retain as a permanent foundation in its subsequent evolution. In the front of the brain there appears the *corpus striatum*, which controls the automatic reactions of emotive life and inhibits the influxes of the cerebral trunk and cerebellum, where the movements of orientation and locomotion are outlined. In the back of the brain, the spinal cord, which once and for all has eliminated the old nervous system, controls the reflex acts that had been absorbed and repressed by the direction of locomotion. Thus the development of the nervous system is presented as a dialectic in which the successive levels of behavior are superimposed on each other – each formation inhibiting the preceding one and absorbing its influxes in a broader regulation in which they are maintained as *suppressed, preserved, transcended*. At each stage, a new moment of consciousness is constituted; in actual behavior, as lived through and acted out, the repressed reaction is *lived* because of the very fact of repression, as an interior movement that has, from that very time, been accomplished, and in which the *ideal life* of consciousness as *intentional synthesis* is constituted. At the lowest level, we find the spinal reflexes where the original reactions of attraction and repulsion, which are outlined in the receptors, are repressed and experienced as a series of impressions that are maintained in the unity of a sensation. The intentionality of *retention* is, to be sure, none other than the repressed movement of attraction or repulsion that is maintained, during the accomplishment of the reflex, as a unity of pulsations that 'still' endure while they are continuously sinking into the immanent past. The reflex act, which develops in the objective form of displacement, is itself only outlined, since the spinal influxes are absorbed immediately in the centers of the cerebral trunk and the cerebellum, which inhibit and integrate them in the more extensive circuits that control orientation and locomotion. The repressed displacements are experienced in an ideal movement of '*protention*', in which the subject becomes conscious of anticipating its immanent future by a collection of directions that outline the horizon of the sensorial field. The influxes where locomotion is outlined are absorbed in their turn by the *corpus striatum*, which inhibits them and transcends them by an automatic movement of apprehension. Repressed locomotion is experienced as an intentional movement that ideally attains to the external object in a *perception of the phantom*. In the evolution that follows the stage of the fish, the influxes of the *corpus striatum* are themselves absorbed in the cortex, in which the forms of manipulation and detour are developed; apprehension is repressed therein and experienced as an intentional movement in which the subject becomes conscious of ideally grasping the object and possessing it in a *perception of the real thing*.

Thus, we see that, at each level, the meaning as it is lived does not reveal the actually real behavior, but rather the outlined and repressed movement of anterior stages. The real act is reflected within consciousness only by the passage to the following stages, where it will be repressed by a new form of behavior that, in its turn, will escape as such from experience. We also see that such a definition accounts immediately for the 'noetic-noematic' correlation that is characteristic of the intentional structure. The repressed act, indeed, necessarily comprises two aspects: the one by which it is outlined and the other by which it is repressed. To the extent that it is outlined and tends towards its goal, it determines the meaning of the intended object *qua* intended (*noema*), defined precisely in terms of this very goal; in this way it is consciousness *of the object*. But to the extent that it finds itself arrested and repressed, it is reflected as the very movement of the outlines wherein intentionality itself consists as constituting synthesis (*noesis*) that is given in the *consciousness of self*. The relation of the constituting or intentional synthesis to the intended meaning as constituted object can be verified at each stage of development. At the lowest level, we find, on the 'noetic' side, retention as the constituting synthesis that 'retains' instantaneous impressions and integrates them in the unity of sensation; on the 'noematic' side, we have sensation 'itself', as an experience that endures in immanent time. Yet it is easy to see that the retentional synthesis is none other than the very flow of rudimentary attractions or repulsions that, *to the extent that they are repressed*, accumulate under repression as an intentional movement wherein consciousness 'still' retains its recently past impressions. Sensation, on its side, is this very flow, but *to the extent that it is outlined* and presented as the unity of an attractive or repulsive meaning that endures as self-identical in lived immanence. At the following level, the protentional synthesis as constituting intentionality and immanent future as constituted meaning are given within the lived horizon of the sensorial field. Yet, the intentionality of protention consists precisely in the act of displacement that, *to the extent that it is repressed*, is reflected as an anticipated movement in which the subject is carried towards its immanent future. The immanent future 'itself', as constituted meaning, is the same act, but *to the extent that it is outlined* and appears in its ideal achievement as this already present future that is outlined on the horizon of the field. With the perception of the phantom, we have, on the one hand, the perceiving act as the constituting synthesis and, on the other hand, the phantom 'itself' as a sensible form that is presented from the outside. But the intentionality of the perceiving act is defined here as the very act of locomotion that, *to the extent that it is repressed*, is experienced as an intentional movement that has, from that very time, attained to its object. The object, on its side, is this very act, but *to the extent that it is outlined*, and thus appears in its ideal achievement as an exteriority endowed with various sensations that are fused into one quality. With the perception of the real thing, which appears at the level of manipulation and detour with mammals, we obtain the same structure: the act of apprehension, *to the extent that it is repressed* by manipulation or detour, is experienced as an anticipated movement in which the subject ideally grasps the object and preserves it in its possession in a perception of the thing; the same act, *to the extent that it is outlined*, appears in its anticipated achievement as the permanently possessed object, or the real and substantial thing. To be sure, it is evident that the original intentionalities of *retention* and *protention* are reproduced at these last two levels as well as at all of those which follow, and that they determine the phenomenological content of the perceiving act in its lived temporality. The movement of pulsations of which repressed behavior consists, is reflected, on the one hand, *to the extent that it is repressed*, in a purely intentional movement that always is retained by constantly transcending itself in the Living Present, and, on the other hand, *in the measure that these pulsations arise* as a succession of states of consciousness that flow on within immanent time.

The distinction, in the repressed act, of the moment of outline and the moment of repression is given immediately in the functioning of the nervous system to the extent that it implies the constant control of the inferior by the superior centers. A proof of this is to be found in the experiments of decerebration that, practiced at different levels, release each time the influxes of the immediately inferior center, whose chaotic manifestations make possible the representation of the moment of the outline – inhibition by the superior center at the normal stage defining the moment of repression. Thus, the release of the *corpus striatum* in the dog by the removal of the cortex reveals excessively violent emotional reactions and a number of purely automatic reactions, which are repressed at the normal stage, by the functioning of the cortical fibers. The automatic apprehension is, indeed, normally inhibited in manipulation or detour. The repressed movement is self-referential in a self-reflection that defines the *consciousness of self*. But the reflection maintains the act of apprehension in the form of an outline by which the subject is conscious of grasping the object and of possessing it as a 'thing': consciousness is conscious of itself as *consciousness of the object*. The same demonstration can be re-enacted for the inferior centers. The release of the cerebral trunk and cerebellum through mesencephalic dissection produces a particular exaggeration of postural tonus and reduces the subject to elementary forms of locomotion. The vertebrate is given over to the anarchic play of its medullar reflexes. These forms of behavior define the form of outlines that, repressed at the normal stage, constitute the lived moments of sensorial life. One sees the profound error of the mechanistic materialist, who defined thought as a secretion of the brain: this would forbid, in principle, every possibility of understanding and would ignore, at the same time, the structure of the actual movement. Indeed, the constitution of consciousness could not be defined simply in terms of the organic plane, since it implies an original dialectic in the course of the nervous influx: more precisely, the structure of the nervous system can be understood solely in terms of the evolution of behavior, which makes us transcend the biological framework as such. In this way we can account for the *meaning* of lived experience, the interiorization of behavior in the form of repressed outlines constituting this intentional relation to the object, in which the living organism is elevated to subjective life. The scientific explanation is not an abstract 'reduction', but rather a genesis that reproduces the actual movement of the *coming to consciousness of material reality*.

The correlations with the development of the child naturally bear solely on the general dialectic of the structures. Nevertheless, they comprise the essential determinations that have just been explained above. The infant still has at birth only reflexes that essentially concern the ingestion of food, and thus refer back to the very origins of the nervous system in the coelenterates, where we see oral behavior appear for the first time. At this level, experience can be defined only as pure *sensation* – i.e., pleasure or pain. But from the very first days, the process of interiorization develops in the form of a disposition to reproduce reflex behavior for itself: the infant begins to suck vacantly at any object whatsoever, notably his thumb if it happens to meet his lips. At the end of four weeks, the form of pure reflex is transcended and absorbed in the behavior of orientation or locomotion: the child puts his thumb in his mouth and systematically sucks it. He directs his eyes towards the light. The capacity of directing movement in this manner implies that sensation is no longer presented as a simple attractive or repulsive given, but as an orientation-value in the *sensorial field* in which the horizon of the immanent future is projected. The form of orientation develops by the progress of postural reactions and the active inspection of the field: the child little by little succeeds in maintaining a sitting position and in keeping his head straight; he looks around and feels with his fingers. The structure of the sensorial field also allows for the formation of the first conditioned forms of behavior. Thus, the child smiles when a human face comes near him; he shows by his behavior that he 'recognizes' his mother, as well as familiar objects like his bottle. Naturally, at this stage the object as such is not yet perceived; the smile of the infant does not mean that he sees his mother yet as an external reality, and even less as a person. He simply reacts to a certain configuration that is outlined in his visual field and that possesses a particular attractive value, seeing that it elicits satisfaction that is indicated by the smile.

The sense of exteriority implies, on the behavioral plane, the possibility of grasping the external object. Yet, apprehension results from a complex development that is not achieved until the sixth month. It is only then that we can say that the child *perceives* an object outside of the lived immanence of the sensorial field. However, this exteriority does not yet imply any reality properly so called, as substantial permanence. Thus the child, who is perfectly capable of taking hold of his bottle, will not be able to turn it over if it is presented to him bottom side up, even after the teat has been made to disappear from his visual field by a progressive movement of rotation. In a more general manner, the child is incapable at this stage of taking hold of an object that is being passed before his eyes behind a screen. The object is but a *phantom*, whose sole existence consists in a lived movement of locomotion in which the subject is conscious of already having ideally reached it as an external place, but not yet *grasped* it as a thing: apprehension defines the actual reality of behavior; it is not yet reflected in the intentionality of consciousness.

4. THE MOVEMENT OF THE EXTERNAL SENSE

With the constitution of the object-phantom in fish, the transition from the internal sense to the external sense is achieved. But exteriority still appears only for the individual object, to the extent that it is detached from the lived immanence of the sensorial field. The horizon of his objectivity as 'world of phantoms' is constituted only with the development of the actual content of the act of apprehension by the transition from life in the water to life on land. In the uniformity of the aquatic habitat, shot through with brief appearances, behavior could hardly go beyond the automatic stage of the emotive reactions of apprehension and flight – flight being the negative mode of apprehension. Thus, fish that are adapted in a unique way to life in the water have evolved very little since primeval times. But those fish that ventured forth on *terra firma*, notably the Crossopterygii, which could use their swim-bladder as a lung and their lobed fins as legs, encountered a multitude of excitations beyond comparison to the rare encounters that occurred in the aquatic habitat. In the face of the plurality of objects to be reached or avoided, the movements of apprehension or flight cancel each other out, and by their reciprocal action, end up in the form of *exploration*. The latter develops in the amphibians and, especially, reptiles with the progress of the cerebral cortex and the formation of the neck, which separates the head from the rest of the body and enables it to turn right and left to the extent that the animal evolves. The movement is constituted in a series of oscillations, which still remain on the level of apprehension, but already go beyond its automatic form, inasmuch as it no longer implies one sole act of absorbed orientation, but rather a simultaneous multiplicity of orientations that are outlined and immediately repressed within the actual act. In such a repression, the subject is conscious of a multiplicity of phantoms that have already been attained and whose whole structure defines the original sense of *space* as the form of exteriority. The object no longer emerges as a simple individual appearance that transcends the lived interiority of the sensorial field: it has its determinate place in the spatial environment and is presented as a moment of the *world of phantoms* in which reptiles crawl. Paralleling this development, the last phase of apprehension (i.e., the absorption of food) takes a new form that once again allows it to go beyond the stage of automatism. In the face of the plurality of present objects, the animal hesitates and suspends its final movement, as can already be observed in the toad that marks time before ensnaring its victim, and especially in the snake that balances its head and carefully aims before striking. During this act of *retarded apprehension*, the object is grasped, already, in a first apprehension that is outlined and immediately repressed, and thus begins to acquire the sense of a 'thing' that is ideally possessed in perception. It is no longer simply the phantom intended as *moving object* in an automatic reaction, but already a reality – an object of *covetousness* in which the repressed act of apprehension is reflected. Nevertheless, such a sense of reality is limited in a strict sense to the lived actuality of *desire* that glows in the eyes of the animal while it delays the capture of its prey.* The tension of desire, up to this point, only marks the transition from the immediacy of the emotion to the permanence of the feeling that will appear in mammals, in which the perception of the actual and substantial thing will be constituted in a definitive manner. However, within the limits of a given activity, desire can be stabilized in behavior in which the object is constituted with a certain degree of permanence: thus, the crocodile keeps its prey dangling before eating it. We can also cite the development of sexual behavior that reveals in various cases, notably in lizards, that the 'other' is perceived no longer as a simple passing phantom, but as a permanent reality: thus, during its sexual period, the grey lizard surrounds itself with a harem of six or seven females which it governs as lord and master. On the other hand, the Rhynchocephalia practice strict monogamy, apparently for life. But, of course, these latter two are extreme cases: the life of reptiles has gone beyond the stage of automatism, but not beyond the very form of apprehension in which the sense of reality can still free itself only imperfectly and sporadically from the pure extension of the world of phantoms.

The perception of the object as substantial reality is effectively constituted only with the behaviors of *manipulation* and *detour* that appear in mammals. *Manipulation* results from the development of retarded apprehension – the forelimbs keeping the object in place while the animal is preparing to put it into its mouth. Thus, there evolves the familiar posture of little animals (i.e., rodents and insectivores) which hold food between their forelegs in

* (See Table of Sensori-Motor Development in the Appendix. – Trans.)

order to eat it. The cat plays with the mouse before killing it. The act of manipulation that, in its original phase, consists in keeping the object at a certain distance from the mouth, represses the outlined movement of apprehension and refers it back to lived experience in the form of an intentional movement in which the subject has already ideally put the object in its mouth, and possesses it in this perception as a real and substantial *thing*. Correlatively, the development of exploration produces the behavior of *detour*. Already at the stage of the pure world of phantoms, the actual structure of the environment has imposed upon the crawling of the reptile the objective form of the *skirting of obstacles*. The interiorization of this form, by means of anticipated outlines that reproduce it at the level of nervous commands, ends with the act of *detour*, which is nothing other than a moving round, that is realized on a previous skirting that has already been outlined. In such a movement, the first moving round is repressed and experienced as an intentional act in which the pursued object is, by that very time, apprehended beyond the obstacle that, ideally, has been skirted. The object, then, is no longer perceived as a phantom that exists only in its actual appearance, but as a real *thing* that subsists permanently behind the screen of the obstacle. Thus, a dog which is shown an object that is thrown through the window does not hesitate to go out the door and down the stairs in order to look for it in the yard. Speaking more generally, carnivores make detours in order to gain advantage, while herbivores adopt the same behavior in order to avoid the carnivores. The acquisition of a sense of reality marks a notable progress in intersubjective relations, which are elevated to the form of *feeling*: the 'other' that is perceived as a permanent substance no longer simply motivates emotions like joy, fear, or anger, but also the corresponding feelings of love and hatred. The permanence of the feeling that makes possible the subjection of mammals to the complex conditioning of domestication accounts equally for their organization at the wilderness stage into families and hordes, in which social life no longer consists in a simple assemblage of individuals, but in the differentiated unity of a collaborative activity. Thus, when wolves hunt in a pack, they divide their labor – some of them getting away from the group in order to bring game to it, sometimes at the risk of not being present to claim their share. In turn, the herds that are attacked often group together in such a way that the females and the young are left at the center, the males remaining at the periphery and facing the assailant. In a more general way, the structure of the feeling defines the very form of subjective life that attains to the perception of the permanent thing. It produces the autonomy of the subject that keeps in mind its past emotions and applies them to the present situation, thus directing itself according to the unity of an objective meaning such as is constituted throughout the multiplicity of appearances. The object is no longer simply an attractive or terrifying form, but a reality that the animal has learned to recognize – thus, a *unity of experience* that elevates behavior to the level of an intelligent and voluntary activity is developed.

This is the stage that is realized in ontogenesis at the approximate age of nine months. Whereas during the preceding period the child was not capable of turning over the bottle that was presented to him in such a way that the teat would disappear little by little, or of taking the rattle placed under his pillow, we now see him searching for the object behind the obstacle. The new form of behavior results from the development of primitive apprehension that once begun, tends to continue even if the object disappears from the visual field. But the realization of *detour* immediately defines a higher level in which direct apprehension is repressed and experienced as the intentional grasping of a permanent substance. At this level, we also observe the appearance of the act of manipulation. Already, at the preceding stage, the scheme of apprehension was prolonged by movements of agitation that produced more or less attraction effects in the external objects: thus, the child who shook the rattle that was suspended at the top of his crib by pulling on a cord, begins to repeat the same movement. The act of shaking is constituted as a new form of behavior that implies, in him, a movement of apprehension that is outlined and immediately repressed, and that refers back to consciousness in the form of an intentional movement in which the effect obtained by agitation begins, already, to be grasped as a thing. Thus sketched out, in a sporadic way, is the sense of an ideal possession that tends to be reproduced each time that an interesting sight stops; and by its very stopping, it awakens an outline of apprehension that is absorbed immediately in the scheme of agitation in which it finds itself repressed and experienced as the intentionality of *desire*. Thus, in an observation of Piaget, a child of seven months, whose laughter has just been elicited by the observer's scratching on a pillow, as soon as the observer stops scratching, starts to draw himself up, swing his head, shake his hands and those of the observer, etc. The development of these immediate forms of behavior to 'prolong the interesting sights' ceases at about the ninth or tenth month with the first forms of *manipulation*. In an experience that is identical to the preceding one, the same child, instead of becoming agitated, shaking, striking, etc., starts to push the hand of the observer in the direction of the pillow. In such a form of behavior, the automatic act of apprehension is repressed completely and experienced as an intentional act by which the subject definitely regards the hand and the pillow as an ideal possession, and thus perceives them no longer as simple external forms or still unstable objects of desire but as actually real *things* that he *relates to one another*. This relation, however, is still presented only on the plane of the pure extension of the horizon of phantoms: it is not constituted as an actually real relation. At this stage, the object has only acquired its sense of reality as an individual object to the extent that it is detached from the world of phantoms as a permanent substance: it is not yet perceived as a moment in the totality of real relations. The 'typical reaction' of this stage consists precisely in the fact that the child, having once found the object behind the screen, searches for it again in the same place, even if one has changed its position under his very eyes by placing it behind another screen. Thus, in another observation of Piaget, a ten-month old child from whom the observer has just taken the rattle in order to hide it under the mattress to the child's left, finds it immediately: but when the observer has taken the object again in order to place it under the same mattress, but this time to the right, the child, who has attentively followed the movement, turns around as soon as the object has disappeared and searches for it to his left. The object has therefore been perceived, indeed, as a substance that continues to exist after its disappearance behind the screen, but its displacement has not been grasped as affecting its sense of reality. In other words, the object does not present itself as an element of the real world in which all existence implies a collection of real relations: it arises from the pure extension of the world of phantoms as a simple, singular reality without relationship to its further displacements.

The horizon of the real world as the place of real relations is constituted only with the development of manipulation in the act of *throwing*, an *intermediary* behavior that makes its appearance in the lower apes. Manipulation, in quadrupeds, still consisted solely in moving the object with the fore-limbs by keeping it a short distance from the mouth that was always ready to grasp it. With the development of tree-life, the fingers become more supple from clinging to branches and finally terminate, in the primates in the formation of the hand – because of which a new form of apprehension is constituted that is of wider scope at the same time that it entails a decisive extension of the field of manipulation. The manipulated object is moved in all directions, in an act that is interiorized under the form of anticipated outlines in which the animal is prepared, from then on to displace the object. The movement that is outlined is summarized by being absorbed in the act of *hurling* or *throwing*. In this new form, the act of manipulation is found repressed and experienced as an intentional movement in which the subject is conscious of carrying the object along its trajectory, and thus of having in his ideal possession not only the object as such, but likewise, the trajectory that it runs through. Thus, the displacements of the object appear in perception as real movements whose combination defines a spatio-temporality and a real causality as the structure of a real world. In such a horizon, the objects enter into real relations that are revealed to consciousness in the act of *mediation*, originally realized in the form of *intermediary* behavior. Indeed, the result is that, in the experience of manipulation, the displacement of an object brings closer another one that was beyond the reach of the subject: thus, the animal, by manipulating a branch, can accidentally bring to itself the fruit hanging at the end of it. Since manipulation has been accomplished in this manner in an act of apprehension, the process is reproduced by anticipated outlines that end up by being constituted in a new form – being summarized and absorbed in a *mediated apprehension*, in which the animal takes hold of the fruit by pulling on the branch. In such a form of behavior, the anterior movement that is outlined is immediately repressed in the form of an intentional movement, in which the subject is conscious of an act of manipulation that is achieved in an apprehension and thus retains, in its ideal possession, the movement of the branch that brings the fruit. Here again, and rather evidently, the reality of his behavior escapes lived experience, since it does not consist in the succession of a manipulation and an apprehension, but rather in the apprehension of the fruit through the *intermediacy* of the branch. In other words, the subject still perceives the branch merely as a real relation of solidarity with the fruit – the displacement of the fruit only becoming the same as that of the branch within the intentional movement of manipulation that ends with apprehension: thus the branch appears to consciousness as a simple extension of the fruit and not as what it is in fact; namely, a distinct object that serves as an intermediary. Hence the form of *mediation* does, indeed, define the reality of behavior; it is not yet interiorized, however, within consciousness.

The act of hurling or throwing, as well as intermediary behavior, appears in the child towards the end of the first year. First of all, the object is let go after a sudden movement of displacement, since the child does not take gravity into account. But the movement is progressively summarized and absorbed in the scheme of hurling or throwing that does, indeed, present itself as the result of the development of manipulation. Similarly, intermediary behavior arises in play in which the child, by pulling in a chance manner on a string, brings the object nearer and apprehends it. But the succession of the two movements is summarized and absorbed in a unique form in which the child takes hold of the object by pulling it towards him with the string. In these new forms of behavior the external environment appears to the subject no longer as a simple extension of the world of phantoms from which a certain number of singular realities stand out, but as a horizon of displacements and real relations that are perceived in a repressed act of manipulation that ideally carries the object through the whole course of its trajectory and draws out the relation marked by the movement of the intermediary that connects with the coveted object. In such a perception, however, the relation is still grasped as actually realized in the present field – the lived act of intentional manipulation being capable of defining only the real relation by which it gives access to apprehension, and not the relation as such in its ideal being as independent of the actual positions of objects. In short, the string is only perceived as a real extension of the coveted object and not as a second object that mediates the movement but would protect its own individuality and would thus be able to be presented in other positions. The intermediary appears only in the actual form of behavior; it is not yet absorbed in lived intentionality.

The sense of mediation is revealed to the subject in the *utilization of the instrument* that is constituted as a development of intermediary behavior in anthropoids,[1] as well as in the child at the approximate age of eighteen months. At the preceding stage, a simple optical contact already sufficed to set in motion the movement of mediation: if the coveted object is presented at the end of a string or of any other object such as, for example, a stick, the subject unhesitatingly pulls on this intermediary. The failure that results from this provokes new attempts in which the subject begins to strike the object with the stick and to move it in all directions. As the displacements end in some fortunate cases in bringing the object nearer, the movement tends to be reproduced by outlines of this form, and ends by being absorbed in a new behavior in which the subject, instead of simply pulling on the stick or shaking it, applies it to the object in a movement that brings the object within reach. In this use of the *instrument*, the anterior movement is repressed and experienced as an intentional movement in which the subject is conscious of getting hold of the object through the intermediacy of the stick. Thus, once again, the real form of behavior escapes consciousness. For, in fact, the stick is not a simple intermediary that is defined in terms of its relation to the present situation, but rather an instrument that functions by its *intrinsic efficacy*: the subject, however, does not perceive it in this precise effectiveness, but simply in the relation in which the movement by which it gets hold of the coveted object is mediated. The instrumental function will be understood only at the following stage, with the employment of the *tool*, in which the mediating object will no longer be used simply in the present situation, but also produced, preserved, cared for; and so it acquires for consciousness the meaning of an efficacious instrument in itself. At the level at which we have arrived with the evolution of anthropoids, the subject, undoubtedly, has already gone beyond the simple perception of the real relation between the intermediary and the object, and begins to rise to an inchoate representation of the ideal relation that he realizes precisely by putting the instrument in its place. But the ideal being so represented is still constituted solely within the limits of the perceptual field. As Köhler remarks, the stick, or any other object serving the same function, must be placed in the vicinity of the end in view, in such a way that both can be perceived at the same time, or at least so that the subject can pass from one to the other by a slight motion of the eyes. Naturally, once the habit is acquired, the chimpanzee begins to look for the stick that it needs. But there again, behavior is determined solely by the immediate situation, since the instrument is not given any attention outside of the occasions of its actual use. The ideality of the relation is therefore still constituted only on the plane of the elementary imagination in which the mediation that is evoked by the present data is outlined and is immediately absorbed and repressed by the real act: the chimpanzee, by getting hold of the stick, no longer perceives it as a simple given reality, but *imagines* it in the ideal relation that mediates the apprehension of the coveted object. In this way, the form of the intermediary is reflected in consciousness, but only in the actuality of its operation. The *image* is precisely none other than this *self-same pure becoming* in which the real transcends its immediate reality in order to be absorbed in a *pure mediating movement*. But the movement is valid only through its actual emergence in the given situation; it is not reflected in the ideal form of its mediating function. The operation of the instrument is imagined in the evanescent experience of an outlined mediation as *the movement of pure passage*: it is not fully represented as a *determination* of its being that is independent of the circumstances of its actualization.

The perception of the instrument *qua* instrument, in its intrinsic efficacy, is constituted only with the use of the tool, which marks the appearance of man. The tool is the culminating point of the dialectic by means of which vertebrates absorbed and assimilated the pure exteriority of the phantom that was presented as a purely external datum in the act of apprehension in fish. The reptile *explores* the ambient space; the mammal *skirts* the obstacle and *manipulates* objects; the lower apes *profit* from the intermediary; the anthropoid *makes use of* the instrument; but man alone can *make use* of the tool. The permanent use of the tool, defined by the behavior of *conservation* and *production*, results from the natural development of the occasional utilization of the instrument. Already in anthropoids, the instrument could assume, through habit, an incipient autonomy and stability. Thus, according to an observation by Guillaume, a chimpanzee that has become accustomed to using a stick refuses to give back the instrument that it has just used and places it securely under its foot. But this act of conservation does not go beyond the short time span that is still influenced by the preceding situation, and, after several minutes, the animal surrenders the stick without resistance. On the other hand, in numerous experiments that are cited by Köhler, we already notice the appearance of an occasional activity of production: thus the chimpanzee Sultan makes itself a stick by breaking off a branch from a tree, by unbending a coiled wire, or by fitting two different bamboo rods together. It even tries to *remodel* a board by biting one of its ends in order to fit it in the opening of the rod so that the necessary length is obtained. Thus, a new form is being constituted in which the simple utilization of the instrument is transcended and repressed, and experienced as an intentional movement in which the subject is conscious of ideally utilizing the instrument which it keeps or prepares and which thus appears to it already in its intrinsic efficacy as an *ideal power* that it holds in its possession. To be sure, these are only extreme cases that sporadically mark the limit of animal intelligence: nevertheless, they bear witness to the process that has been realized on a vast scale in phylogenesis, when at the beginning of the quaternary period, the receding of the forest imposed a return to the soil, where a multitude of new problems favored the intellectual development that had begun in tree-life. The scarcity of available vegetation obliged the human ancestor to return to an omnivorous diet and so spurred him to make permanent use of the stone and the stick in order to kill game. The anthropoid stage was transcended because of long experience, at the end of which man had acquired the habit of carefully guarding the instruments that he found in nature; and, since he was in need of producing them, he broke a branch off a tree or broke a stone in such a way as to make them more prehensible. Such behavior presupposes nothing more than the development and stabilization of the occasional use of the instrument just as it is presented in the anthropoid: nevertheless it constitutes a radically new structure in which the form of utilization is definitively absorbed and repressed and experienced in constantly renewed outlines that make the tool that the subject produces and keeps with him appear to him as a *potentially powerful* object. The ideal relation of mediation is presented no longer as a simple, evanescent *image* that arises on the occasion of a given situation as a purely disappearing moment; it begins to be objectified in the relatively stable unity of a *representation*, properly speaking, in which the mediating operation is presented as a practically acquired possibility in consciousness that is always available with its intrinsic efficacy. However, here again, the reality of behavior escapes lived intentionality. For the act of production consists in a *labor* of which the subject has not yet become conscious as such, since, by making the tool, he represents it primordially to himself in the simple function of its possible utilization, and not in its objective structure as it is constituted in the articulations of this very labor. The movement of labor as *productive force* is reflected in experience only through the behavior of *organization* and *education*, in which the process of production becomes the object of a conscious will in the primordial form of the will of the *other*, as *another self*.

The forms of organization and education do not comprise, to be sure, any absolute creation: they result necessarily from the development of the elementary acts of *tradition* and *collaboration*, such as already appear at the anthropoid level. Thus, in Köhler's experiments, we see the chimpanzee Sultan, forbidden by the observers to participate in the experiment, showing a less experienced companion how to lengthen a stick by fitting two bamboo rods together. In other cases, the subjects unite their efforts in putting a box under bait that has been suspended from the ceiling, so that it can serve as a stool. Certainly, it is not yet a matter of intentionally social activities, since the relation between the different subjects refers, in fact, only to the objective universality of the instrumental function inasmuch as it provokes the same reactions in all the animals that are present:[2] nevertheless, it is still the case that the development of the instrument that leads to the tool is necessarily elaborated within the framework of communal activity, and so engenders the use of the tool in the form of collective habits, as techniques that are acquired within the group and transmitted from generation to generation. Hence, the immediate forms of behavior, collaboration and tradition, are differentiated by being articulated in these techniques and, in this way, elevated to the form of *language*. In the practice of language, the agreement between individuals is no longer derived simply from the objective universality of the instrumental function, but is constituted in a movement in which the subject is conscious of intentionally realizing a communal work and a transmission of power: the meaning intended by the verb is precisely none other than the very operation of production. The extent to which it is outlined and experienced as an ideal work in which this power of mediation that the subject, in the actual use of the tool, represented to himself as a simple functional determination and that he now posits in its objective structure by the act of *judgment*, is defined on the plane of the objectivity of the concept. But the intentionality of language presents certain peculiarities that compel us to retrace our steps in order to examine the relation of the *expression* to its *signification* in its general movement.

The universal notion of expression or symbolic function includes the totality of the existent in the sense in which we speak of the 'expression of a landscape'. But we obviously understand 'expression' here in its strict sense as intentional expression that consciously intends a signification. From this point of view, it seems to be presented for the first time in mammals. Even though the production of sounds is already encountered in certain fish, it can only be a matter, at this level, of a simple accompaniment of action itself. The case of amphibians and reptiles would offer matter for discussion, notably in regard to sexual appeal, but we prefer not to consider these still doubtful examples. On the other hand, it would be difficult to deny the presence of intentional signification in the barking of a dog opposing the entrance of an unknown visitor, or in the behavior by which it manifests its desire to play at returning a thrown object, or to lead its master towards a definite place. One could not very well speak, here, of a simple accompaniment of action, since the very action in question (e.g., to chase the stranger away or to return the ball) is not really accomplished, but simply sketched out, and is only valid *qua* sketched out in this barking or this coming and going in which the animal has the air of being precipitated toward an imaginary object. The sketch of the act appears as the precise *expression* that symbolizes the act itself that has been left unfinished. Thus, we can define the intentional relation of signification in pure terms of behavior: the symbolic act of signification is an act that stops at the initial phase of its accomplishment and refers, thereby, to its total form that has been left unaccomplished. The intended signification is precisely only the unrealized part of the behavior that is outlined and set in motion by the sketched out movement. Here again, it is not originally consciousness that determines behavior, but rather behavior that produces consciousness. The appearance of the visitor releases in the animal a sudden movement that is at once stopped and repressed by the presence of an obstacle or by reason of previous conditioning. Thus, the reaction is only sketched out by the barking, and it is achieved by a simple outline of attack that is not actually realized, but which thereby gives to the subject the consciousness of *signifying* its intention of driving away the stranger. The latter, on his part, responds by sketching out a movement of flight, but without really accomplishing it, since the animal itself had only sketched its movement of attack. The flight thus is sketched out, and just as soon arrested, is achieved in the form of an outline that is experienced as an ideal movement in which the subject perceives the behavior of the animal as, to be sure, an attack, but as an attack that is simply signified, but not actually realized. In other words, the barking of the dog appears to the visitor as an expression whose meaning he *understands*. Thus, since the phenomenon of expression has been defined as a sketched-out behavior, understanding results immediately from the response, since it itself consists only in a sketching out that is repressed at once. We should note that this sketch, in its turn, functions as an expression whose signification refers back precisely to understanding itself: by sketching his movement of flight, the visitor *signifies* that he has understood the behavior of the animal. In its turn, the animal responds by redoubling its barking, demonstrating thereby that it understands that the other has understood it and should be prepared to flee. In this way the primordial correlation, in which the foundation of reciprocity is defined, is constituted. Since the symbolic act, or the sketched-out act, provokes a response that is itself sketched out and in which is expressed the manner by which the other is affected by the expression of the first, it is this being affected itself which, in its turn, affects the first subject and arouses in it a new expression. Thus, the real movement of sketched out behavior is experienced as an exchange of intentional acts of reciprocal understanding, each affecting the other in the very same manner in which he himself is affected: the other affects me by means of the affect by which I have affected him.

It should be noted that we are clearly distinguishing the two notions of *sketch* (*ébauche*) and *outline* (*esquisse*). Outline stops at the level of nervous commands, and so is reflected as intentional movement. Sketch is, on the contrary, a real act that begins at the muscular level, but cannot be accomplished because of an external obstacle or previously ingrained conditionings, or even because the stimulus is itself but sketched out and does not reach the threshold necessary for a complete response. Yet, the sketched out act implies precisely its very accomplishment in the form of a repressed outline that is experienced as an intentional movement in which there is defined the *signification* whose sketch becomes *expression*. In this way a new form of intentionality is constituted, a form in which the intended meaning is no longer simply experienced in the subjectivity of consciousness, but appears in behavior itself as the meaning that it expresses. From this, the subject derives the feeling of having gone beyond the real content of his acts through the *intention* that is manifested in them, and which thus seems to define them, giving their 'true meaning'. But such a 'going beyond' is only an appearance that conceals the actual reality of the movement. The intention *qua* intention does not yet refer back to the truth of behavior, but simply to its unfinished aspect, which is only explained in terms of strictly objective reasons: it is but an empty outline, an imaginary complement of the action that is unable to reach its goal. In fact, here again it is not consciousness that goes beyond the real, but rather the real that goes beyond consciousness. Indeed, the symbolic act *qua* behavior is constituted with an actual content that escapes, once more, the subjectivity of lived experience: for from the very fact of its signifying function, it is no longer at the level of the form that it expresses, but is necessarily elevated to the level that is immediately higher. Thus, to consider the *real meaning* of behavior, the dog that barks chases the stranger away through the *intermediacy* of its barking: but the *intended meaning* in its consciousness is, evidently, concerned only with the very act of chasing the stranger away, which symbolically refers back to the scheme of *manipulation* and not to the barking *qua* intermediary. It is thus true that the act, sketched out, has this singular privilege of transmitting to experience *qua* signified intention the form of behavior that ordinarily escapes it: but this is precisely the case only in order to go beyond it and absorb it in a higher structure.

The symbolic function is constituted in mammals, in its roughest form, as a simple result of the interruption of the real act. It thus appears only in a sporadic manner within the framework of direct behavior. Its conscious utilization is produced only at the level of the instrument: with anthropoids, where one sees the behavior of *simulation* being developed. Everyone knows that chimpanzees in captivity love to play tricks on visitors by pretending to give them an object and quickly withdrawing it as soon as someone extends his hand in order to grasp it. We then see them enjoying the discomfiture of the naive visitor, and sometimes even profiting from the situation by hitting him or stealing something from him. Here, the symbolic act no longer functions simply as an immediate expression that implies its own meaning, but already as a *sign* utilized by the subject in an intention that is not confused necessarily with the one it intends to express. In other words, the movement of expression has developed in the form of an *instrument* that the chimpanzee uses in order to draw the visitor to it as with a stick. We see immediately that the repeated performance of such behavior ends by stabilizing it into a permanent deployment, in which the instrumental function of the sign appears as constantly available: thus, the chimpanzee that is on the verge of being caught in the act assumes an innocent look and feigns being busy with something else. Hence, vocal emissions can already take on the symbolic signification of an efficacious mediation. Despite the insufficiencies of positive research in this domain, it seems that the anthropoids, outside of the immediate expressions such as are generally found in mammals, use a certain number of sounds that enable them to understand each other with relative precision, at least in what they have to say, and which, in principle, must refer back, not of course to concepts, but to general forms of mediation as modes of the instrumental function. Thus, in an observation by Crawford, we see a chimpanzee calling out in cries to a friend in order to help it pull a heavy box by means of a rope. At any rate, we can attempt to clarify the problem by referring to the verbal schemes that develop in the child's second year before the appearance of the first phrases. Naturally, we should be very cautious here, since the emission of articulated sounds appears very early in the child, so that we cannot very well speak, in this domain, of a parallelism that is comparable to the one that is observed in the sensori-motor development. It is impossible, however, not to notice that the vocal emissions of the child, before functioning as words that designate conceptual structures, refer to concrete means of acting upon the object, and so appear as symbolic forms of instrumental efficacy. Thus, in the observation of Piaget, we see in a child of eighteen months the vocal sign 'Panana' designating not only the grandfather, but also everything else that is wanted by the child, even in the absence of the grandfather – he is, to be sure, the most efficacious instrument of the child's desires. Another says 'Mama' when showing his father a lamp to be turned on and off, even though this game, as the observer specifies, is an exclusively paternal specialty. 'No more' signifies a departure and the throwing of an object to the ground, and it is applied to an object that turns over and to the game of giving an object so that it can be returned: thus it would be a matter here of a symbolic mediation that is no longer aimed at obtaining the object, but at removing it. The emission of the vocal sign is presented as a sketched out act of utilization, ending in an outline that ideally mediates the desired effect. We readily see the difference that separates these instrumental significations from the conceptual forms that will appear with language, properly speaking, towards the end of the second year. The subject still intends only general modes of efficacious mediation: he does not have the concept of determinate operations in which the symbolic structures are defined by the objective form of a productive *labor*. It is true that with the use of verbal schemes, real behavior already begins, in fact, to be raised to the level of the tool: and the constitution of the vocal sign is presented precisely as a first form of the process of fabrication. But the act of production still appears only in the actual content of the movement; it has not been interiorized within a lived signification.

We now have the elements that are necessary to re-embark upon our development and to define, in an adequate way, the intentionality of language. Inasmuch as the symbolic act is a sketched out act whose meaning consists in its very outlined and repressed accomplishment, the utterance of the verb results from rhythmic halts in the collective use of the tool, as a *sketched out movement of production*, ideally ending in the productive operation itself, and experienced as a *signified intention*. The ideal work of the concept is but the very movement of real work being interrupted for an instant by virtue of its objective structure and being continued on the symbolic plane by the use of the word. In other words, it is true that man has spoken because he 'has something to say'. But what he 'had to say' was not presented originally in an intentional form: the human ancestor did not say what he thought because he thought it, but thought it because he said it, and he said it because he stopped doing it.[3]

It is time to specify the universal notion of production as it results from the use of tools. We have seen it being constituted in the form of a collective activity in which the occasional preparation of the instrument is stabilized and developed at the anthropoid stage. The chimpanzee breaks off a branch from a tree in order to use it as a stick, but such a movement intends only the present situation, and the instrument is abandoned as soon as it is used. Yet, it is clear that the permanent use of the tool has entailed, from the emergence of the very first human horde, the constitution of a number of collective habits as techniques that are acquired in the group and which do not solely concern the fabrication of each tool, but also the modalities of its use. This use is presented as a collection of definite operations which, by their operating structure, *produce* the desired effect. Thus, through the function of the tool, the form of production extends to the totality of human activities, inasmuch as they are henceforth objectively determined by the technical habits of the group. When the human ancestor hurls his stick to kill game, his gesture is no longer defined simply in terms of an instrumental efficacity, but rather in terms of an intrinsic structure that is established precisely by virtue of the material structure of the permanent tool. The same is true for the behavior that originally concerned only the body itself, such as the acts of walking, running, climbing, etc., as they are integrated as determinate moments in the general process of production. In short, if we define productive activity as activity that ends with its result according to *rules*, it is clear that these rules are constituted, first of all, outside of consciousness, in the reality of behavior as an objective consequence of the material conditions of the use of the tool. But in collective life, such determinations are reflected on the symbolic plane by being sketched out in sounds which, from this very fact, take on an equally determinate form: the animal cry becomes human language by being articulated within the structure of productive labor.

Thus, the *ideal* production of the object, in the positing of the judgment, finds its authentic foundation in the *real* production, where articulated sounds are engendered. Universality is but a *result* in which is reflected the indefinite possibility of repetition that is implied in the objective structure of the process of the tool. Thereby, the phenomenon of knowledge, in its historical upheaval and in its truth-value, is ultimately justified. The classical theory of knowledge, proceeding by simple reflection on the judgment, created an impassable gulf between the actual reality of consciousness that returned to a pure 'psychology' and its truth-meaning that was a carry-over from 'transcendental philosophy'. The real existence of the object disappeared, at the same time, in a formal notion of objectivity that systematically reduced it to the simple condition of an ideal agreement between all possible subjects. The primordial meaning of truth as 'the adequation of the intellect and things' became, thereby, totally unintelligible. The whole philosophical effort after Kant consisted in returning to the 'concrete' as the identity of being and meaning. But the idealistic prejudices remained immanent in all of these attempts that regularly consisted in a simple juggling of the real in a more or less subtle 'interiority', in which the horror of bourgeois thought for the harsh materiality of productive labor was covered over. Correlatively, mechanistic materialism took upon itself the task of demonstrating, through the very evidence of its own absurdities, the impossibility of 'reducing' consciousness to matter. *A priori*, it was forbidden to oneself to understand the real dialectic in which meaning is constituted as the result of the natural movement of being. Dialectical materialism provides the authentic realization of this 'return to the concrete' which, according to bourgeois thinkers, is maintained obstinately at the stage of *simple pretension*. Henceforth, the primordial familiarity between man and nature is reestablished, but at a higher level in which the mediations that constitute the grandeur of classical rationalism are preserved and absorbed. The material genesis of thought does not suppress the meaning that it intends; it justifies it in its true intentions as the meaning *of* reality.

Naturally, at the stage at which we have arrived, we have still constituted only the general form of the concept in its *possible* agreement with things. The actual adequation depends on the power of techniques that are actually elaborated. Mankind, at its beginnings, had at its disposal, in almost all cases, a purely symbolic mastery by means of the simple exercise of language, which worked an imaginary transposition of the elementary productive schemes upon all the perceptive givens. Thus, children who are beginning to speak identify clouds with the smoke of a pipe, the wind with the air blown by the mouth, growing babies with air bubbles or any object that can be inflated. One of Piaget's subjects, seeing the waves moving back and forth over small strands of sand on a beach, exclaims: "It looks like a little girl's hair being combed." Whatever the inadequacy of these judgments, the form of *production* is acquired as the authentic 'condition of possibility' for a true knowledge. All further progress will consist in transmitting the ideal work of the concept from the purely symbolic stage of verbal exercise to the rational stage of logical operations, reproducing the complex organization that assures the efficacy of real labor. But the development of production implies the dialectic of social forms, in which human life is constituted in its properly human meaning.

5. REMARKS ON THE PRECEDING DEVELOPMENT: THE PASSAGE TO THE DIALECTIC OF HUMAN SOCIETIES

It has been a long time since the animal origin of man has seriously been contested by biology. However, there remained the task of posing the problem on the plane of consciousness, or of dissolving the mystery of the relation of self to self by determining the genesis of intentional significations in the evolution of species. To be sure, from this point of view, we had to consider only the major line of evolution that led to man. For this reason, we have neglected groupings that otherwise would be of considerable interest, such as of invertebrates that followed after worms or birds. On the other hand, it is clear that the observations that we have mentioned are valuable only by virtue of serving as examples that illustrate the general levels of development. Species of animals that exist at the present time are too 'specialized' to be capable of being made part of the ancestors of man. The forms upon which evolution took place have disappeared because of the very plasticity that allowed them to evolve. Nevertheless, it is appropriate to try to form an idea of them through the actual representations of evolutionary branches that remained stationary: what is essential is the obtaining of a lived comprehension of the evolutionary movement, whose reality and principal stages are indubitable. We could still be reproached for not having discussed the biological processes in detail. But these are not our immediate concern, and it is sufficient for us to have shown the identity of real and lived structures, since the course of the nervous influx implies a series of differentiated circuits in which consist the outlined and repressed forms of behavior by which we have defined intentional significations. Moreover, we have admitted that the advancements within the nervous system depend upon exercise within the framework of a specific experience, since the heredity of acquired characteristics has been demonstrated by the admirable works of the Michurinist school. At any rate and whatever the problems that are still confronting the biologist, our genesis of the structures of consciousness remains valid, since it only refers directly to the very fact of behavioral evolution.

In regard to the determination of intentional significations, the superabundant literature of the philosophers should provide us with ready-made frameworks. The concrete analyses here are, unfortunately, rather rare and constantly obscured by metaphysical contradictions. Only the works of Husserl have provided us with actual help, at least in regard to the constitution of the primordial 'thing'. In regard to the genesis of judgment starting from the 'thing', the phenomenological description provided decisive results for the elimination of classical prejudices, but we would search there in vain for precision concerning the concrete content of actual mediations. The objective method alone, by means of the distinction of forms of ***intermediary, instrument, tool***, and ***language***, permitted the explication of the phenomenological articulations of the lived movement that elevates the real given to the ideality of the ***true***. From sensorial impressions to the perception of the 'thing', the forms of outlined and repressed behavior solely concern the movements of the body proper and can be found again, without excessive deformations, in lived interiority. Starting from the 'thing', the activity of the subject implies an external mediation whose reflection in consciousness becomes enveloped in a confused penumbra, and, despite the unique mastery acquired by Husserl in this domain, intentional analysis hardly went beyond the level of general indications for future research. Though extraordinarily rich, if compared with the formalism of previous theories, they are revealed as disconcertingly poor with respect to the actual content of the real development.

To the extent that the forms of production will become more complicated with the development of human societies, the lived structures will become less and less accessible to the direct investigation of the experience of self-consciousness. And, in fact, the phenomenology of ideal forms has never gone beyond the stage of abstract distinctions between 'essences', whose content it remained incapable of defining. Contrary to a widespread prejudice, the privileged domain of pure phenomenology does not concern human significations, but rather primitive and properly animal levels: sensation, the sensorial field, the object-phantom, and the primordial 'thing'. It is here that the descriptive technique, once divested of its idealistic gangue, makes possible a decisive confrontation between consciousness and behavior in which it is conclusively established that there is nothing more in the intentional movement than a real movement that is outlined and repressed. Inasmuch as repression is performed by behavior itself to the extent that it is effectively achieved or, in the case of symbolic acts, by other objective givens, it is patently clear that lived experience is only an abstract moment in the dialectic of the living body acting as a whole.

Naturally, we have not, thereby, by any means, taken away any role from consciousness: such a claim would even be self-contradictory, for since matter is defined in terms of movement, consciousness necessarily acts as a material movement. All that we are contesting is the concept of a pure consciousness whose relation to the world would imply a metaphysical transcendence. But in the immanence of natural being, consciousness plays a major role for the accomplishment of behavior. This point is evident with regard to symbolic acts, which acquire a new content on the plane of the real precisely because of his very symbolism. But even in cases in which the new structure appears for the first time in the real act as the necessary result of the previous development, since the lived meaning is produced only by the inhibition performed by this act on the preceding level, it is clear that, for the rest of the movement, the repressed outline, in its turn, supports the behavior that is accomplished within it. In other words, *if it is action that, originally, gives birth to lived experience, then lived experience, in its turn, prolongs and supports action by its repeated outlines*. We can even say that at each level, once progress has been achieved by the dialect of behavior, consciousness *directs* action, even though the real form of action constantly escapes it. The notion of direction does not imply the mastery of real behavior, but simply the mastery of the levels previously established. Thus, it is correct to say that apprehension is normally directed by the perception of the phantom; but such a relation simply means that the subject is conscious of dominating in advance the place intended in a repressed outline of locomotion; while, in fact, the act of apprehension does not grasp a simple 'place', but rather a real and substantial 'thing'. But the latter appears to consciousness solely at the subsequent stage with the act of manipulation. In the case of symbolic behavior, the role of consciousness is more important, since it is the outlined or symbolic achievement of the sketched out act that gives it the value of an expression, and thus elevates real behavior to a higher form. But this form itself escapes lived intentionality. Thus, in the case of the child who does not yet know how to construct phrases, the verbal sign is already the object of true production and so takes the form of a tool defined by an objectively determined structure, even though its intentional meaning bears only on its mode of effectiveness for its instrumental function. In short, the idealistic reduction of ***being*** to ***being perceived*** is valid only for the levels already transcended. In the actuality of his present life, the subject necessarily determines himself in terms of the real itself *qua* external to the intentionalities of consciousness: *consciousness directs behavior but does not determine it.*

However, with the appearance of productive activity and its reflection in the symbolism of language, a reversal occurs in which behavior seems to come under the effective control of consciousness. During the whole course of animal development, the subject only progressively anticipated the ***given*** from the instantaneousness of the impression to the still immediate permanence of the 'thing', that is hardly transcended in the instrumental behavior by an evanescent image of mediation. With the movement of labor, the given is absorbed in the ***product*** and is posited as such by consciousness in the act of judgment. Henceforth, the subject seems to be able to rediscover himself in things, and the constitutive dialectic of the human species is presented as a ***passage to freedom***. However, it is still only a question of a first degree that disappears in the actual content of development. For the very movement of production necessarily engages the subjects in a number of human relations whose real structure, once again, escapes their consciousness. In fact, as long as the progress of the world market and major industry have not integrated the totality of producers, by means of the power of technology, within the unity of a common task, the various enterprises remain constituted piece-meal on the foundation of natural givens, and necessarily encroach upon one another in their development. The immediate form of collaboration that we have seen engendered within the original experience of labor ends, paradoxically, in a radical opposition between the different unities of production and thereby imposes upon them a universal behavior of hoarding: the covetousness that rends economic life reflects the power of expansion of the general process that is in each, to the extent that it has not yet reached the saturation point where it will be organized into a unique system. However, each moment of production implies a factual solidarity with several others, and the technical conditions of real labor resolve the conflicts within a collection of more or less provisional coalitions, inside of which the struggle is restrained by a game of *exchanges*, a game which expresses the actual relation of the opposing forces. The modalities of exchange define the *productive relations* (i.e., the human relations such as those that objectively result from the movement of the *productive forces*) and are imposed as the form of equilibrium that defines, for each, the maximum development that is compatible with the totality of producers.

It is here that there is presented this extraordinary movement in which consciousness purports to be elevated to a pure spirituality by being alienated from itself in the phantasmagoria of a total mystification. All signification that we have seen constituted up to this point remained immanent to natural life: the very labor of production creates the human world only as a moment of nature, and the inner autonomy of judgment merely expresses the objective structure of this real production. The ideality of consciousness, so understood, does not consist in denying, but rather in affirming on the plane of lived experience, the reality of material existence. Yet, within the dialectic that we now have available for examination, consciousness is posited in an unexpected manner as the *suppression* of all external reality. The exchanges that we have just seen engendered on the economic plane, by virtue of the real relation of the productive forces, suddenly take on the form of a *negation* of this very same reality: they appear to consciousness as *not* being motivated by economic necessities, but by *pure relations of right* within the horizon of a *transcendence* that, from that time, already excludes all exteriority as such. But if we carefully examine the movement, we notice that it contains nothing more than the symbolic reproduction of real acts of exclusion and elimination that developed within the economic conflicts and constantly reappear within the ever unstable equilibrium of productive relations. Indeed, so long as the unification of all the productive forces has not been attained, the form of hoarding is maintained as the actual intention of the game of exchanges, and the subjects continue the struggle with sketched out acts ending in an outline that is experienced as an intentional movement, in which each one is conscious, from then on, of ideally eliminating his competitors, at least to the extent that they would encroach upon the forms of equivalence which guarantee to him his own mode of acquisition. In other words, until the event of the realization of the proletariat and the socialist relations in production, every form of *possession* implies the *exclusion* of the totality of producers and so creates the consciousness of a pure interiority as *exclusive* of the materiality of real relations, inasmuch as this materiality implies precisely the objective possibility of a participation by everyone. Undoubtedly, nature as such has been transcended, already, by the labor movement, and the judgment has posited the meaning of the object as human product: but such a product preserved its reality-value within the horizon of natural existence, of which human life was but the apex. With the passage to the economic relations of the production or exchange relations, the form of possession terminates by enclosing each within the limits of his own development and in ideally constituting him in a *pure consciousness of self*, denying all exteriority as such, since reality itself is admitted, henceforth, only within the horizon of a possible possession. Thus, the being of the object takes on, for consciousness, the paradoxical signification of a *negation* of actually real objectivity, since from that time the subject has already absorbed its being in his lived-through *project*, and acknowledges it as existent only to the extent that it conforms to the ideal conditions of this project in which are reflected the real conditions of his actual hoarding.

In this way the movement of transcendence is constituted in which movement this *expropriation* of the totality of producers is achieved and which constitutes for everyone the establishment of *private property*: the passage from the fact of possession to the right of ownership of property is accomplished by this symbolic act of universal exclusion in which the subject is always ready to take up the struggle in order to defend his exclusivity. Consciousness, however, remains absorbed in the ideal side of its movement. By denying the exteriority of the object, consciousness has posited itself in the absoluteness of the Self, without realizing that this Self is but the Self of the *proprietor*. The destruction suffered involuntarily in the course of economic struggles appears afterwards, in the interiority of memory, as a sacrifice that gives to the resulting economic relations the absolute value of *sacred* relations. So, having reached the human stage, consciousness *conceals* from itself its own origins by being absorbed within a *horizon* that defines for it every truth-meaning, but nevertheless reflects the limits of the mode of production of its material life. The ideal appropriation of the world in the spiritualism of self-consciousness is but the symbolic exaltation of the social forms of real appropriation, and the phenomenological dialectic of the *spiritual forms* is the movement of mystical appearances that cover over the truly real history of the defense of private property.

In the social production of their life [says Marx in *Contribution to the Critique of Political Economy*], men enter into definite relations that are indispensable and independent of their will, relations of production which correspond to a definite stage of development of their material productive forces. The sum total of these relations of production constitutes the economic structure of society, the real foundation on which rises a legal and political superstructure and to which correspond definite forms of social consciousness. The mode of production of material life conditions the social, political and intellectual life process in general. It is not the consciousness of men that determines their being, but, on the contrary, their social being that determines their consciousness.[1] *

CHAPTER TWO

THE DIALECTIC OF HUMAN SOCIETIES AS THE BECOMING OF REASON

6. USE-VALUE AND THE MOVEMENT OF SACRIFICE

During the nomadic life of mankind at its very origins, the productive forces that were constituted with the evolution of the tool could not yet assume the form of wealth; the house (which will later guarantee effective separation and the conditions for accumulation) had not yet appeared. However, as far as we can tell from ethnographic documents, pure use-value as such already entailed an *appropriation* whose form was modeled, evidently, on the very conditions of the use of the appropriate object. Thus, in the societies of hunters that we know of, the individual was consistently acknowledged as the owner of his weapons and tools, while the hunting ground, whose actual occupation was possible only on the collective level, belonged to the group. And so it appears that the sense of property derives quite naturally from the very exercise of productive forces: the appropriation of use-values is constituted with the appearance of an absolute legitimacy. Nevertheless, property implies in fact an *expropriation* of the totality of producers, since it does not consist in the simple enjoyment of the object, but rather in the *de jure* exclusion of every participation by others, while it belongs precisely to the real being of use-value to be available to all. To be sure, the opposition, at this stage, between individual ownership of weapons and tools and the collective ownership of the hunting ground is itself internal to the general form of private property, since it is still only a question in the second case of *private groups* that defend their exclusivity. Thus, it is here that there occurs the primordial mystification that will justify the whole later process of exploitation: for exclusivity that is attached to the possession of the object entails the *negation* of its actual reality and its absorption within the transcendence of a pure consciousness of self, in which the *mystical union* of the proprietor and his property is accomplished.

If the refined structure of the great religions of history poses problems of interpretation that are sometimes delicate, primitive religion naively reveals its content in the immediate practical character of its functions. It guarantees to the group the exclusive right of occupying its territory by the constitution of the sacred powers of the natural environs, at the same time that it sanctions individual property by attributing to the human subject a mystical essence. Thus, according to observations of the Australian aborigines, made by Spencer and Gillen, there is never any attempt at annexation or encroachment either among the different tribes, or between the local groups that make up each tribe

> yet, this is the result of the very firmly-fixed belief of the natives that they are the direct descendants or reincarnations of ancient ancestors, who were especially associated with certain local spots at which their spirit parts remained when they died and their bodies went into the earth.[1]
>
> Still further examination of each local group reveals the fact that it is composed largely, but not entirely, of individuals who describe themselves by the name of some one animal or plant. Thus, there will be one area which belongs to a group of men who call themselves kangaroo men, another belongs to emu men . . . .[2]

Thus, even though the evolution of the clan has led to its subdivision into several groups, each of which can admit strangers into its territory, nevertheless, it is still the case that the totem is retained as the center of the spiritual life of the local group. And there is no doubt whatever that this mystical kinship is the basis for ownership of land, since the essence of the cult consists precisely in commemorating the manner in which the totemic ancestors came to occupy, in the legendary times of *alcheringa* (the 'dream' time) the land that their descendants now possess. On the other hand, we know that, among all primitives, the individual ownership of useful objects also presents a magico-religious character that is particularly noticeable in the very wide-spread custom of burning or burying the weapons and tools that belonged to a dead person. Here again, myths that attribute to man a supernatural origin have the obvious function of bestowing on the real producer a sacred power that confirms him in his rights as a proprietor.

Naturally, there could be no question of making religion into an artificial invention that is consciously designed for the justification of ownership. The spiritual appropriation of the world in beliefs and rites achieves, in a necessary manner, the dialectic of real appropriation, such as it results from the movement immanent in the mode of production. This is what is demonstrated in the examination of the religious content itself. Every religion, through an infinity of diverse traditions, is constituted in a symbolic movement in which the subject *suppresses* himself in his natural existence in order to be elevated to the consciousness of his 'true being'. *Sacrifice*, prior to being an offering to a god, is the *becoming of the sacred itself*, in which the real is transcended through its self-annihilation. Thus, even though the totem is not yet a sufficiently individualized being for one to be able to offer gifts and prayers to it, the Australian aboriginal rites almost always consist of a consecration that involves human blood. More particularly, in the ceremonies of initiation, the adolescent finds himself being inflicted with extremely painful mutilations, such as the extraction of a tooth, circumcision, and subincision (all of which symbolize the death of his natural being and his access to mystical life). Yet it is clear that a symbolism of this kind can refer only to violent struggles that arose with the appearance of the first use-values constituted in productive activity and that are retained as an absorbed and ever-renewed moment within the equilibrium of social relations that are defined by the status of property:[3] the young initiate *becomes* a 'true' man with the acquired rights of the members of the group through the symbolic repetition of the *actually real becoming* in which there was constituted, through bloody conflicts, a form of appropriation that is recognized by all, since it sanctions the productive relations or exchange relations that are asserted in the very movement of productive forces. At the original stage at which we place ourselves, there is not yet any issue regarding exchanges of goods, which would obviously supposes an already acquired sense of property, but simply regarding exchanges of services that are implied in the objective conditions of the use of tools; a collaboration is asserted within each group for the collective hunts and the manufacture of equipment, and, likewise, relations of good neighbors are established between various groups, owing to the necessity for everyone to be dedicated to the search for food in the absence of objective conditions of exploitation. Thus, the primordial behavior of monopoly is *repressed* by the form of common interests, the exchange of services obliging everyone to respect the use-values possessed by others, since they are precisely necessary for the execution of these very services. Thereby, an actual possession that is transformed by the development of its immanent content into legal property is practically assured. In fact, the conflicts thus mollified are not entirely suppressed, but merely *interiorized*: the movement of exclusion that is characteristic of monopoly assumes the form of a *repressed outline* that is experienced as an intentional movement in which the subject constantly endeavors to exclude others, in an ideal manner, from the enjoyment of the object. It is not simply a matter of use-values that he already possesses and appropriates in a definitive manner by the affirmation of a systematic exclusivity. The act of hoarding is pursued on the intentional plane for the very objects that are possessed by others, and the *recognition* of the other's possession, which transforms it into property, is precisely possible only on the foundation of an outlined exclusion that is experienced as *negation*. Indeed, to acknowledge an object as the property of another is already to be, at that time, in possession of it in order to renounce it immediately and to ascribe it to the other. In other words, the movement of monopolizing is experienced as immediately arrested, but the very arrest is precisely possible only because of the gesture that has already been outlined; and, indeed, I claim, by returning the object, to give it a meaning that it would not have without my recognizing it. Thus, the social constitution of the right of property originally implies for everyone a universal movement of appropriation that is equivalent to a universal *expropriation*. The idealism of legal relations of right is constituted precisely in this process of exclusion that is constantly outlined, in which I *deny* the actual reality of the object inasmuch as it is or can be found in the possession of another, in such a way that at the very moment in which I attribute it to this other I still make his right depend on my own act of consent. In such a movement of *transcendence*, the object is perceived as *not being* this real being that is given in the materiality of practical life, but as a pure ideal essence in mystical communion with pure lived subjectivity. Animals and plants are no longer these natural realities that are actually possessed by man through the use of the tool; they are presented as pure supernatural forms that, from this time, are already available in the interiority of a pure spiritual self-consciousness. Without such a *horizon* of magical relations, the subject recognizes himself in one of those forms which define *for him* all sense of 'truth'; he *is* the kangaroo, the emu, or the hakea flower. Through the objectificaton of the act of appropriation in the abstraction of the mystical substance, totemism completes the alienation of the producer by confirming him in his powers *qua* proprietor. *Sacrifice* reproduces, on the symbolic plane, the actually real drama in which the genesis of the state of right through the original violence is presented to consciousness under the mysterious appearances of a *death* of the natural being that is followed by a *resurrection* in the form of the sacred essence: for the monopolizer who has risked death in defense of hoarded goods is reborn as a proprietor. Religion is not an external sanction of property: it is its intimate exaltation and passionate justification.

We understand now why primitives, about whom all the evidence agrees especially in emphasizing good, practical sense and acuteness of observation, seem to live in a world that is perfectly foreign to real experience. Thus, while they made use of their weapons with skill and shrewdness, they remained convinced that the efficacy of these weapons was due to a supernatural power, *mana*, which they felt obliged to develop through magical rites that were extraordinarily painful and often cruel. It is clear that aberrations as universal and persistent as these could not be explained simply through ignorance. Moreover, when we examine the actual content of the notion of *mana*, we discover, instead, an authentic beginning of knowledge, in the form of elementary schemes which define for the primitive how to use his objects. Thus, the *mana* of a spear is, in fact, only the very power of this spear to kill the adversary through the action of the point. More generally, legendary beings, in their fantastic exploits, make use of the same kinds of weapons and tools as real men in their daily life: the knife, the spear, or the boomerang. And if they also use supernatural instruments, it is only a question, if we consider the effects, of symbolic substitutes for profane instruments. In a word, mystical action reproduces the structure of real activities, and since it has been demonstrated that the *concept* is only the symbolic form in which a determinate moment of production is outlined, the magical object, as imaginary object, is defined as the immediate application to sensorial givens of conceptual schemes that are acquired in the experience of productive labor: it is the *symbolic tool* whose ideal use, somehow or other, finds its foundation in the efficacy of the real tool upon which it has been modeled. We do not thereby depart from the movement that is immanent in material life; the opening of consciousness to the world in which the truth of being is revealed is but the flowering of productive activity that is constituted, through the symbolism of language, into a universal structure of production. And yet it is this very revelation that is presented as a mystery in which the true meaning of being is, paradoxically, transmuted into a *negation* of the actually real being – real determination being denied and *concealed* in the supernatural form of a pure mystical efficacy: the spear does not kill because it is pointed, but because it is *mana*. In fact, economic development itself involves a negative moment: the act of hoarding and exclusion that *conceals* the actual reality of the object and camouflages it from the ambitions of the other. *Revelation* as *concealment* defines the primordial mystification in which the real being of the producer is transcended into the sacred essence of the proprietor in a mystical identity with his property. However, within this concealment, the consciousness of ever-possible conflicts is preserved, and the being of the sacred appears as infinitely terrible; for in the depths of self-consciousness it refers to the actual reality of property as exclusive and always threatened enjoyment.

Thus, mystical life is not confined to the simple lived experience of mystery as such. In the face of sacred terrors, which reflect the real dangers of reciprocal exclusion, mystical life is organized into a system of supernatural relations that preserve the subject from these mysterious dangers by sanctioning the real relation of forces in the exchanges imposed by activities of production. Since the whole of the natural world has been transferred to the mystical plane by the movement of monopoly that has been outlined and repressed, the equilibrium of real forces between the various producers comes, thereby, to be reproduced as a form of appropriation that is valid for all. Such a form is extended, evidently, to the totality of being, since the object is recognized as existent *qua appropriated*. Thus the various groups of the tribe, by each appropriating part of the territory, also share the natural beings: plants and animals, the sun, the moon, the seasons, the wind are distributed among the various totems that symbolize, for each group, its power of appropriation.[4] Religion, as a general theory of social relations, defines simultaneously a 'popular logic' in which the form of appropriation of economic goods assumes the value of a *category* that is the foundation for the classification of concepts, and the *a priori* condition for the comprehension of the world. In this way a network of regulated participations is constituted, which network prevents possible conflicts and preserves in everyone the primordial movement of hoarding. For this very order is elaborated in the substance of the sacred, in which everyone is conscious, from that very time, of appropriating the totality of being, since that very same thing that escapes him as belonging to another is posited as such solely by his own operation. By transposing to the supernatural plane the real movement of reciprocal concessions, the state of right accomplishes this marvel of restoring the sense of exteriority in the very form of exclusion that is experienced as negation and transcendence: thus, union is realized in distinction itself, and the subject is elated by the intimate certitude of an absolute freedom in which everyone rediscovers himself in the other-being by *constituting* the latter within himself.

Such is the mystical joy which, with the privative form of appropriation, justifies, for consciousness, the alienation of real existence in the transcendence of mystical essences. An ecstasy of this kind implies, rather evidently, a positive content that is denied and concealed, but that nevertheless founds the authentic meaning of spiritual existence The enjoyment of the *Self qua universal Self* does, indeed, present itself in its real form in the primordial moment of human relations or productive relations in which each producer truly *transcends* himself in the universality of social labor. Thereby, *conceptual truth*, originally defined in terms of the technical structure of production, is elevated to the fullness of its meaning as *the truth of existence*. In the movement of cooperation the subject extends his own activity by *symbolic outlines* that embrace the totality of the communal task and make him become conscious of universality as the true meaning of his singular existence. The speech (*le verbe*) that emerges from the depths of organic life accomplishes the unity of the *self* and the *other*, and the form of exchanges that represses the hoarding attempts is imposed on everyone to the extent that it enables everyone to realize himself in the plenitude of his productive forces. Thus, the *transcendence* of individual subjectivity is accomplished within the *immanence* of natural existence. The subject, however, is conscious of it only in the supernatural form of a mystical transcendence. Indeed, as long as the development of production has not brought about the suppression of the privative form of appropriation in a fully humanized nature in which a guarantee is provided for everyone according to his needs, the act of hoarding will be maintained through the equilibrium of exchanges and, in constantly renewed outlines, will absorb the material reality of the object in the exclusivity of a purely spiritualistic consciousness of self. Egoism is not an eternal trait of any 'human nature' at all; it is derived as a necessary result from the inorganic form of production, which polarizes the various producers in permanent conflicts, in the very progress of their productive activity. And the apparent infinity of human desires is but the reflection of the actual movement of total production within the particularistic horizon of independent producers. Hence, conflicts will truly be suppressed only through the systematic unification of all of the productive forces, which, when in full bloom, will guarantee the elimination of the real basis for the process of reciprocal exclusions. Until that time, egoism will be able to be repressed in certain forms of cooperation: of necessity, it will be preserved *qua repressed*, and repression appears in the form of a limitation within the spiritualistic horizon of total hoarding. Thus, the real joy of the producer in the movement of exchange relations that allow for the development of his productive activities is transformed, paradoxically, into a pure mystical exaltation, which denied the actual reality of these very relations and arbitrarily places them under the spiritual dependence of the *Self* – the monopolizer abandoning only a part of his claims in order to save, ideally, their universal form. Yet, under the fallacious appearances of an absolute autonomy, the expansion of singular existence, in its actual transcendence towards the social whole, is reversed in the submission of this very same singularity to the tyranny of transcendent forms, and the expansion of the natural being in communal work is perverted in the morbid enjoyment of a pure *renunciation*, in which the spiritualism of pure self-consciousness persists in the behavior of hoarding by means of the obstinate symbolism of a powerless exclusivity.

To the extent that the level of productive forces is elevated, human relations are broadened and impose an even richer truth-content on the primitive form of the sacred essence. At the primordial stage of nomadic economy, the relations of cooperation place each producer in an immediate unity with his means of production, and the recognition of property integrates him within the very being of ownership as his permanent status. Consciousness is identified with the object within a horizon of mystical participations that enclose human relations within a narrow network of rigid obligations. Marriage is nothing but a remarkable form of *statutory property*, such as is constituted in the exchange of services. However, within the framework of the family, new productive forces develop that determine a first liberation of the human subject: domestic economy and the beginnings of accumulation give rise to an exchange of goods, in which the proprietor is detached, in practice, from his property and thus asserts himself in his properly human existence. However, the evolution of wealth profits only a small number, and the independence of the person, such as comes to be constituted in real relations, takes for the subject the mystical form of pure *prestige* in which the preservation of the act of hoarding under its ideal negation terminates in the establishment of property of the nobility. At the primitive stage of statutory right, everyone disposed, in practice, of the same means of production, and magical power still remained very little individualized. With the concentration of wealth, there appear personal spirits, which justify the fortune of the chiefs. However, within the framework of feudal exploitation, agriculture and crafts were developed, which activities required simplification of the circulation of wealth through the mediation of money. The exchange of commodities serves to liberate human relations from the natural determinations of blood ties, in which the feudal stage still enclosed them. Production, thereby, is no longer limited to the narrow perspectives of use-value: it is organized in terms of the *exchange value*, which objectively determines the form of a *rational universality*. However, mercantile economy is still profitable only at a particular level, separate from the mass of workers, and, in its turn, is forced to accede to the position of the dominant class. Henceforth, the rational structures elaborated in the new mode of production are themselves alienated, in traditional form, from the sacred essence, and the barbaric might of the primitive god is elevated to the predicates of *justice* and *wisdom*, symbolizing the compromise imposed by the bourgeoisie on the nobility, in the form of the universal power of the monarch as the guarantor of *contractual right*. Thus, the movement is presented to consciousness solely in its upside-down form: it is not a question *for it* of a new alienation that permits the perpetuation, in the guise of transcendence, of the privative characteristic of the appropriation of use-value through the universal content of exchange value; it is much more the materiality of productive labor that appears to consciousness as 'alienation', and the establishment of *mercantile property* is experienced by it as a 'return upon itself'. As an exploiting, but at the same time oppressed, class, the bourgeoisie can be opposed to feudal arbitrariness only in the name of universal values that are authentically constituted in real production, but it accepts this very same universality only within a spiritualistic horizon that preserves its own possibilities of hoarding and enables it, once recognized, to associate with the ancient ruling class for communal defense against the general structure of oppression. 'Alienated' in his forced alliance with the mass of workers, the member of the bourgeoisie 'rediscovers himself' in the exclusivity of his property rights and is elated at the idea of 'reconciling', in the infinity of divine wisdom, the rationalism of mercantile production with the mysticism of immediate participation inherited from the primitive human feelings of magician-hunters.

However, within the travesty imposed by the form of appropriation, the real content of the universal develops with the progress of productive forces. In the experience of material labor, new mediations constantly appear – first in the occasional form of the *instrument*, then, through the play of habit, in the permanent structure of the *tool*. From there arise new *practical concepts* that are at first interpreted within the reigning ideology of magico-religious forms, or, at a more advanced stage, as pure abstract idealities which, however, progressively impose their actual truth-content to the degree that the development of production is revealed as incompatible with the exchange relations defined by the established form of property; then a new mode of trade is constituted as an evolution of properly human relations, in which all true value is engendered to the extent that it raises the individual to a superior form of universality, and so liberates him once again from local and familial particularisms in which the mystical participations of the barbaric state are perpetuated. However, as long as the dispersed character of productive forces is maintained, the clash of private interests leads the class that effectively profits from the new relations once more to restrain the new values within the traditional framework of privative appropriation as a new appearance of the spirit, which is obliged to obtain the adhesion of the old ruling classes by preserving the essential features of their acquired rights. Considering the immediate protagonists, we find that history was the movement of struggle and compromises between the bourgeoisie and nobility – the first imposing upon the second broader and broader forms of appropriation, while safeguarding the particularistic basis that conditions its own perspectives of hoarding. However, with the absorption of the labor force in the process of capital and the expansion of major industry, the bourgeoisie has exhausted its possibilities of progressive development. The conflict of the productive forces with the relations of established production can no longer be resolved by a broadening of privative appropriation, but rather by its suppression. Evolution is accomplished within the very framework of the bourgeois system, through the formation of a class whose objective conditions of existence realize the suppression of every class privilege. As a specific product of capitalistic exploitation, the proletariat can accede to property only by suppressing its privative form, and its spiritual life implies the suppression of all spiritualism. To be sure, its struggle which evolves within the element of its ancient society, still implies an ultimate form of *sacrifice*. But this sacrifice is self-conscious. The proletariat *knows* its very self as a defense of class interests, whose objective movement necessarily entails the annihilation of all class structure. Thereby is constituted the horizon in which the real foundation of the 'world of the spirit' is discovered, and history is no longer the history of the managing classes (which is none other, in fact, than a prehistory), but the history of productive labor and human relations. Beneath the mystery of transcendence, in which the behavior of hoarding is outlined, traditional values are revealed in their true substance as forms for the expansion of human existence, such as it is elaborated in the materiality of real life. Dialectical materialism is the dialectic of the proletariat inasmuch as it repossesses, in favor of productive humanity, human significations that have been distorted in their human meaning by the spiritualistic coalition of private interests. In the actual universality of communistic society, the inherited symbolisms of the mysticism of privative appropriation are retained as *pure aesthetic mementoes* of a past that is forever gone: in this way, the alienation that stripped from man the enjoyment of his real life comes to an end, and consciousness itself is recognized in its natural existence, where all being acquires for itself an infinite value; for it *is* nature itself in its *becoming-human.*

At a certain stage of their development [says Marx], the material productive forces of society come in conflict with the existing relations of production, or – what is but a legal expression for the same thing – with the property relations within which they have been at work hitherto. From forms of development of the productive forces, these relations turn into their fetters. Then begins an epoch of social revolution. With the change of the economic foundation, the entire, immense superstructure is more or less rapidly transformed. In considering such transformations, a distinction should always be made between the material transformation of the economic conditions of production, which can be determined with the precision of natural science, and the legal, political, religious, aesthetic or philosophic – in short, ideological forms in which men become conscious of this conflict and fight it out. Just as our opinion of an individual is not based on what he thinks of himself, so we cannot judge of such a period of transformation by its own consciousness; on the contrary, this consciousness must be explained rather from the contradictions of material life, from the existing conflict between the social productive forces and the relations of production. No social order ever perishes before all the productive forces for which there is room in it have developed, and new, higher relations of production never appear before the material conditions of their existence have matured in the womb of the old society itself. Therefore mankind always sets itself only such tasks as it can solve; since, looking at the matter more closely, it will always be found that the task itself arises only when the material conditions for its solution already exist or are at least in the process of formation. In broad outline, Asiatic, ancient, feudal, and modern bourgeois modes of production can be designated as progressive epochs in the economic formation of society. The bourgeois relations of production are the last antagonistic form of the social process of production – antagonistic not in the sense of individual antagonism, but of one arising from the social conditions of the life of individuals; at the same time the productive forces developing in the womb of bourgeois society create the material conditions for the solution of that antagonism. This social formation brings, therefore, the pre-history of human society to a close.[5*]

7. THE MOVEMENT OF WEALTH AND THE BECOMING OF THE GODS

If it is true that terror was the origin of the gods, this terror was not a terror of natural phenomena. For, as weak as the primitive is when he is confronted by surrounding material forces, these forces themselves could arouse solely a passing emotion each time. And, in fact, the religious worship of massive natural phenomena appeared only at a relatively later period of time, when the progress of productive forces required a symbolism adequate to the new horizon of universality. At the primitive stage, mystical anguish is revealed in all its original naiveté as properly *human* anguish: it is the fear of witchcraft, which epitomizes only the fear that man has of his fellow man, since everyone is endowed with a mystical power that makes him a potential sorcerer. It is a universal belief accepted by primitives that death is never due to natural causes, even when these appear in a perfectly obvious way. On the contrary, they attribute it, with an absolute conviction, to an act of magic, and they feel obliged to search out its author in order to punish him. Such obstinate representations imply, also, an *a priori* that can refer solely to the reality of human relations. The terror of the sorcerer symbolizes the threat of aggression that weighs upon everyone in a society in which the use of violence still remains visible below the superficial repression of statutory conventions. If we consider human life from the perspective of its productive activity, its problems are seen as purely positive ones, even when they are not resolved and there is nothing therein to justify the anguish of mystery. But the sense of property awakens, in the depths of the soul, a strange disquietude: it is the primordial *care* in which the *transcendence* of the project of hoarding is constantly brought back to the actuality of the existence of the hoarder as solitude and *dereliction*. Thus, consciousness experiences the need of being reassured through the symbolism of sacrifice, in which the privative form of appropriation appears to it as the magical preservation of its spiritual intimacy.

At the nomadic stage of the hunting economy, however, the elementary structure of production practically assured a factual equilibrium between the various producers, and a relatively simple system of mystical representations sufficed to sanction it on the plane of right. Undoubtedly, the participation that was constituted in the exchange of services already allowed for the first form of exploitation, to the advantage of the older men who, because of their experience, became directors of the collective operations by realizing their influence in a magico-religious superiority that guaranteed real advantages to themselves. Thus, the Australian old men, because of the ritual system, reserved for themselves the youngest women and the best pieces of meat from game killed in hunting. But it is only a question, there, of objects for immediate consumption, and we cannot yet speak of an opposition of classes, since equality is maintained for the essential means of production. With the beginnings of accumulation in domestic production, the movement of wealth explodes the rigid framework of primitive status, unchaining covetousness and installing *war* as a permanent state. The structure of conflict triumphs in all human relations, and the exchange of goods, imposed by the development of productive forces, itself assumes the agonistic form of an exchange of challenges. As a struggle of pure prestige, the *potlatch* is the symbolic re-enactment of actually real struggles that continue to be sketched out under the unstable equilibrium of relations of production. The gift, or, in extreme cases, the voluntary destruction of wealth, sometimes even of the very person of the proprietor, defines the new form of sacrifice, where everyone is given over to a symbolic act of violence that ideally enables him to take possession of the goods of his adversary. In the spiritualism of such a movement, the sacred essence, which was still diffused within the whole group at the preceding stage, is concentrated in a small number of individuals who are favored by the actual conditions of production and exchange. The chief is the one who was able to become wealthy by guaranteeing the subsistence of his companions; whereas slavery marks the incapacity to face the conflicts that define the actual reality of property relations. But the movement is concealed under the magic of the notion of *prestige*, and the form of inequality is fixed in titles of nobility, in which it finds its mystical justification.

It would be difficult to find a more convincing example than the *potlatch* to illustrate the hypocrisy of spiritualistic self-consciousness. Under the flaunted ardor of generous acts, each thinks, obviously, only of enriching himself by obliging his adversary to respond by an even greater generosity, and the whole *art* of this commerce consists precisely in appreciating, rather well, the real possibilities of the partner. But these sordid calculations that betray themselves in the extreme formalism of reciprocal proceedings, are draped in the *transcendence* of a pure negation in which each intention is presented in the form of its contrary. However, the movement is too intensely lived not to condition some sense of truth. And, in fact, the independence of the person in respect of his possessions, wherein the concept of *honor* consists, is realized in an authentic manner in the economic process of the exchange of wealth, in which the producer detaches himself from the particular product by elevating himself to the perspectives of total production. To the extent that such a totality has not yet found the real mediations that will allow it to be determined in an autonomous structure, it implies neither the *abstract* rationality of justice nor the *concrete* fervor of charity, but simply the *immediate* transcendence of particular possessions in the pure feeling of the dignity of the human person.[1] Thus, spiritual values do, indeed, have their authentic foundation and their truth-content in the movement of properly human relations such as they are constituted in the primordial experience of economic life. But this content itself is negated in a mystical form that marks the *status quo* in the game of exchanges of the primitive structure of hoarding. Already, at the preceding stage, the real relations of solidarity, which were constituted with the exchange of services in the peregrinations of the group, were transformed, by virtue of the privative form of appropriation, into pure magico-religious relations, and the authentic meaning of *duty* was alienated in the mysticism of statutory 'prestations' that permitted old men to be guaranteed sensual privileges. With the constitution of wealth, a new progress is attained, in which, henceforth, the human social essence implies the autonomy of various producers who, by elevating themselves to the total movement through the exchange of their own productions, constitute this totality as the meaning arising from their particular existence, and thus transcending the constraint of *duty* in the liberty of *honor*. But such a signification, by virtue of the weak level of productive forces, is immediately absorbed in the exclusivity of privative appropriation, and the collective expansion of particular persons in the movement of exchange is alienated in the *transcendence* of a pure *prestige* that the adversaries dispute among themselves in a veritable deluge of boasts, insults, and reciprocal challenges.

Thus, the possessor of wealth has become a proprietor, and his open disinterestedness is revealed as the price paid for the recognition of the right to hoard. In the lived experience of such a movement, the still little individualized totems of primitive clans are determined in a more elevated form: they are the *personal spirits* that justify feudal privileges. More precisely, *alienation* in wealth is presented in the form of a *possession* by the spirit that defines for the feudal proprietor the intimate sense of his lived subjectivity; it is the mask that covers him with the symbolism of gifts and voluntary acts of destruction and which from then on, preserves him from threats of aggression. However, the form of the divine is still not fully constituted: for it implies a transcendence that absorbs not only the particular determinations of individual existence, but this very same individuality as such. A movement of this kind is realized only with the concentration of feudal power imposed by the progress of wealth. Indeed, the productive forces that develop with agriculture, cattle-rearing, and the beginnings of crafts, give an increasing importance to profane trade, which goes beyond the ritualistic formalism of prestige trade. Yet, in the movement of commodity exchanges, a new structure of human relations is constituted, in which structure the individual transcends himself in a rational universality. But such a signification immediately falls back into the persistent structure of hoarding: it appears as a mystical predicate of royal might, in which the nobility has mollified its quarrels of prestige in order to dominate the mass of workers and absorb, in feudal property the first results of mercantile production. Under the protection of the manorial organization, local trade develops to the profit of the lord, and the king is the delegate of the feudal coalition, the symbolic proprietor of universal wealth. In its exalted form, he is the God who protects human exchanges and takes the benefits of them for himself. In this way there appears the classical notion of transcendence, in which the elementary mysticism of magical participations is enriched by human virtues that are engendered by the new mode of production. At this original stage, however, moral attributes are still little challenged: God imposes Himself first of all by His invincible power as universal mediation in the mode of immediacy.

We understand now the fundamental ambiguity of the concept of divinity and one that has not failed to engender philosophical reflection. For this being, infinitely transcending the world of men, seems to reveal a conspicuous weakness for their prayers and offerings, and while his faithful place all of their hopes in Him, He Himself only lives truly in the movement of their sacrifices. This is so because the meaning of the divine does, indeed, essentially comprise two moments: on the one hand, the universal content where man recognizes himself in his sense of *truth*; on the other hand, the form of exclusivity that paradoxically subjects the enjoyment of this universality to a particular *favor*. Such a contradiction betrays the hoarding of the new productive forces within the framework of feudal barbarism – the universal product *qua* exchange value becoming the object of an exclusive enjoyment that defines the very substance of the divine. The mystery of transcendence is nothing more than the confirmation of the contradictions of privative appropriation, and the supernatural perfection of God refers back to the real universality of commodities, alienated in the irrationalism of exclusivity as it is constituted through the ministry of sacrifice. However, at the level at which we have arrived, the determinations of mercantile society are not yet fully acquired: they are announced only in the organization of the power of chiefs around the royal person, the *sacrifice* of whom periodically allows for the legitimation of the hoarding of new riches. More specifically, the ritual murder of the king defines the very genesis of the divine, the mystical enjoyment of a total appropriation through the symbolic evocation of the universal movement of violence, in which the unstable equilibrium of lordly relations is constituted.

Evidently, the moment of exclusion is, for consciousness, confused with the very content of the object. In the lived experience of transcendence, the subject represents to himself not the exclusivity of his act of appropriation, but the appropriated being itself in the form of its negation. Thus, the sacrifice of the king appears not as a symbolic hoarding of scocial wealth, but as a renewal of the mystical power that claims to engender it. And so rituals are interpreted from the point of view of nature, while it is obviously a matter of a reactivation of human relations. Indeed, if it was first a matter of guaranteeing the abundance of the harvest, or the well-being of the cattle, we would understand, if worse comes to worst, that in the absence of truly efficacious processes, desire did trigger symbolic attitudes. But nothing would justify the sense of negation that is implied in the sacrificial act. In other words, we readily see that the exaltation of the royal person can represent an intensification of productive forces; it still remains to be explained how such an exaltation is obtained precisely by an act of destruction. The problem is that, in the precise exclusivity of privative appropriation, the 'truth' of the object is understood only as a negation of its reality. Thus, the abundance of the harvest is not intended in the immanence of its material being, as a product that is useful for all, but rather in the transcendence of an exclusive enjoyment by the ruling class. The sacrifice of the royal person symbolizes the sacrifice of the nobility, which allows it to hoard the universal content that emerges with the beginnings of mercantile production: the appropriation of commodities is inaugurated as an extension of feudal property, symbolized by the coming-to-be of fantastic spirits in which the omnipotence of the gods is engendered.

8. MERCANTILE ECONOMY AND THE SACRIFICE OF THE SAVIOR, GOD

The constitution of the seignorial regime completes the division of society into classes which, until the triumph of socialism, defines the permanent setting of human history. The hierarchy of chiefs and companions that still preserve, at the tribal stage, some appearances of democracy, is organized into a pure structure of oppression, inasmuch as this hierarchy guarantees the unity of the ruling class against the mass of real producers. From the latter are derived successive levels of the rising bourgeoisie, which imposes its own mode of exploitation in the form of the abstract universalism of contractual right that is realized within the very framework of feudalism by the passage to absolute monarchy. On the plane of consciousness, the movement is presented in a dual aspect. On the one hand, the unification of feudal power is exalted in the unification of the system of gods, which tend to be absorbed into a universal god, the *Sun*, as the symbol of the sovereign power of the monarch. On the other hand, inasmuch as the absolute unity only covers up the expansion of the bourgeoisie, it entails a transformation in the signification of the divine itself. It is no longer simply the proprietor himself who is ready to risk his existence in defense of his goods: it is a matter, this time, of a *universal* sacrifice, since the property to be insured consists precisely in a universal object, a *commodity* that exists as such only in the universal movement of commerce. At the preceding stage, exchanges were limited to the local market within the framework of the seignorial manor, and commodities were still concealed under the form of wealth whose appropriation rested directly on the feudal lord's actual capabilities of defense. With the development of trade and the progress of urban life, property, despite its privative form, involves a universal participation, and all are interested in the security of everyone else, since the circulation of commodities, in which its very being consists, has become so extended that the right of each is already, from that time, the right of all. Hence, to the extent that the specific content of mercantile production is made explicit, the act of exclusion that is maintained in the form of appropriation finds its exalted confirmation no longer simply in the *sacrifice of man*, where private defense is still symbolized within the limits of the seignorial manor – the king merely unifying the feudal forces – but rather in the *sacrifice of God* Himself as symbol of a universal defense of the public order, in which the whole of society pledges its existence for the protection of the well-being of each. The king is now the representative of the bourgeois revolution, and God, in whom he exalts his power, is no longer simply the God who receives sacrifices, the supernatural being of feudal property to whom one accedes by risking one's own life; He is the God who dies and is reborn glorified, the supernatural being of mercantile property, whereby the act of hoarding is justified for each by the sacrifice of all.

Consciousness, evidently, ignores the real content of its lived intentions: the movement, *for it*, presents itself as a simple succession of natural determinations. Thus, the cults of Osiris, Tammuz, or Adonis (where one celebrates the death of God and His triumphant resurrection) are interpreted as a representation of the agricultural cycle of the fall of grain and its burial, followed by its germination. As a matter of fact, Egyptian and Asiatic civilization, in which these divinities were the object of a passionate adoration, owed their prosperity to the mercantile economy that was sanctioned by contractual right. The myth thus refers not to the natural or technical process of vegetation, but rather to the appropriation of its product *as commodity*. From this point of view, the legend of Osiris deserves particular attention, for the *divine drama* reflects therein, with rare fidelity, the vicissitudes of *human drama* in which the struggle of the Egyptian bourgeoisie against feudalism ended with the confirmation of bourgeois property in absolute monarchy.

Osiris, sovereign of the Delta, god of corn and personification of the Good, was murdered by his brother Set, god of evil and of the sterile desert, and master of Upper Egypt. With the help of 72 confederates, Set treacherously enclosed him in a wooden chest which, thrown into the Nile, floated to Byblos on the Syrian coast. The body of Osiris was miraculously enveloped by a tree that preserved it until the arrival of his spouse, the Goddess Isis. She brought him back to Bouto at the mouth of the Nile, where his body was discovered by Set, who tore it into fourteen pieces. Isis discovered them and succeeded in reconstituting the body of her divine spouse. Osiris was restored to life under a double form: on the one hand, as king of the dead, on the other hand, in the person of his son Horus, miraculously begotten in the neighborhood of Bouto, in the swamps of Chemnis, where Isis was hidden. Horus defeated Set in combat, after which the uncle attacked his nephew before the tribunal of the gods. This tribunal proclaimed Osiris 'justified' and Horus his legitimate heir. Henceforth, Osiris was to be the god who awarded immortality in the beyond, while Horus symbolized the legitimacy of monarchical power in this world.

The myth clearly shows the essence of the movement of class struggle that marked the origins of history in Egypt, when the feudal regime was revealed as an intolerable obstacle for the activity of the bourgeoisie.[1] They had appeared in the Delta, during the fifth millennium owing to the development of export trade, notably with Byblos, where the Egyptian corn was exchanged for Lebanese wood. In the mercantile cities a new concept of right was born, a concept which freed the individual for commercial operations by giving to the expression of his will the value of a universal determination, while in the country, rigid structures of feudalism were perpetuated. But the closed organization of the seignorial manor impeded the purchase and transport of grain, which activities were indispensable for exporting it. The struggle for the freedom of exchange and the extension of contractual right to the country brought about the unification of the towns in a confederation that was placed very naturally under the protection of Osiris. Osiris did, indeed, have the characteristic determinations of the trade of Byblos: god of corn, he was represented by the sign of *Djed*, a pruned tree in which one can recognize a Lebanese cypress. For the bourgeoisie of the Delta, Osiris was the god of free exchange, and he personified the Good, in opposition to Set, the master of Upper Egypt, where the seignorial regime retained all its force, despite the influence of the northern merchants who had already established trade branches there. Set was thus the god of feudal barbarism, symbolized by the sterile desert. The plot fomented by Set and his 72 accomplices against Osiris, represented the feudal reaction in which the lords of the Delta formed a coalition with the barbarians of the South in order to attack the confederation of the cities. This confederation was defeated, and its rulers fled to Byblos, where they were received by the guild of wood, which the myth recalls in the episode of the tree that enveloped the body of the god. However, the abuses of the government of the lords provoked popular disenchantment, symbolized by the lamentations of the goddess Isis in search of her spouse, and the Osirians exiled in Byblos were able to return to Bouto, the center of the bourgeois party. The movement was once again dispersed by the feudalists, but was reconstituted clandestinely with the help of the priesthood, which, under the cover of the monarchy, sought to substitute itself for the seignorial nobility: it was, indeed, through all sorts of magical practices that Isis reconstituted the body of the god and resurrected him, and it was equally through magic that she conceived Horus from the substance of Osiris in the neighborhood of Bouto. The defeat of Set by Horus represents the victory in the Delta of the absolute monarchy of Bouto, followed soon after by the unification of Upper Egypt by the kings in Nekken, supported by the mercantile colonies established by the northern cities. Thus, the hymn to Osiris concludes: "Because of His laws, abundance has been established; the roads are free; the ways are open."

We understand now the tremendous popularity of the *Passion* of Osiris along the Nile. It offered to the Egyptian consciousness an exact mirror of the real dialectic that ended with the regime of the Pharaohs, in the sole mode of appropriation that was possible, given the state of the opposing forces. Monarchical absolutism in Egypt justified itself as the only power capable of breaking the seignorial opposition and guaranteeing the public order necessary for the mercantile economy that, by virtue of the expansion that it provoked, is presented as *the true meaning of existence*. In the revolutionary struggle that led to such a regime, the bourgeoisie, to be sure, placed itself at the forefront; for at stake was its very own mode of production, experienced by it as *the authentic mode of its lived experience*, inasmuch as it found in it the *practically possible* form of an indefinite expansion. And since such a universality can be appropriated only on the plane of an organization that is itself universal, the *sacrifice* of the bourgeoisie for the establishment of mercantile property can take its whole meaning only within the horizon of a *divine sacrifice* as symbol of a universal agreement, where the sacrifice of each is the answer to the sacrifice of all. Osiris, "master of acclamations in the nome Anedjeti," is the expression of popular monarchy, and his death is the death accepted by the bourgeois revolutionary who dies, not like the feudalist, for his simple personal presige, but for the defense of a universal order in which the world is defined in its *true being*. However, the Egyptian bourgeoisie was still only developing on a national basis, where it could not find the necessary resources to be imposed effectively as a ruling class. The confederation of the towns was beaten down and dissolved by the seignorial coalition, and the Osirian party was resurrected only by virtue of the help of the clergy, symbolized by Isis, the expert magician, and this represented a stage that evolved from landowners whose interests were already linked to the export trade. The compromise, placed under the sign of Horus, secured the triumph of absolute monarchy that guaranteed the security of the bourgeoisie by destroying its political independence and subjecting mercantile economy itself to a progressively stricter regulation. From then on, the revolutionary ideal represented by the Osirian concept of a democratic royalty was transferred to the 'beyond' where immortality conferred by the savior god achieved, on the symbolic plane, the appropriation of universal values created by the rationalism immanent in mercantile production.

When we think of the merciless exploitation attested to by the funeral monuments of ancient Egypt, we cannot help but admire these boastings in which the dead person engages in a veritable moral *potlach* with the divinity: "I have given food to those who were hungry and clothes to those who were naked. I was a father to the orphan, a husband to the widow, a shelter against the wind to those who were cold. I was one of those whose words was good. I have earned my substance in justice." The intemperate spiritualism of Egyptian religion marked the *practical* impotence of the bourgeoisie of the Nile to make the specific structure of bourgeois property predominate over the feudal horizon that was perpetuated with the absolutism of divine right, sustained by the national economy. The monarch remained the proprietor *par excellence*, and the rational content that appeared with the new mode of production could be appropriated only through the royal person now assimilated to the supreme divinity, the Sun, as the absolute self-consciousness that engenders everything by thinking it. Thus, the Pharaoh, while remaining Horus, son of the popular king Osiris, identified himself, paradoxically, with Atoum-Ra, the universal creator from whom Osiris himself is a true descendant. In such a contradiction, the absolute monarchy transcended its own origins and absorbed within its power the bourgeoisie from whom it itself came. But an exclusivity as systematic as this found its immediate sanction in the concerns of real existence and, in the first place, in the obsession with *death* in which, within the existent relations, the absurdity of mystical pretensions to a universal appropriation is brutally laid bare. And so the remembrance of the authentic moment returns: the popular revolution in which, under the sign of Osiris, the primordial meaning of universal forms is affirmed. Thus, Osiris will be the god who guarantees the fate of the soul in the beyond, the symbol of *constituting movement* whose social reality is concealed behind the mysticisim of transcendence that justifies the form of hoarding.

Thus, if immortality is accorded, in principle, only to those who have shown themselves *just* in this world, this *justice* first of all refers only to the lived form of the appropriation of *goods*. Nevertheless, such a value implies a truth-meaning that finds its authentic foundation in the experience of human relations, such as they are constituted in the movement of productive forces. Indeed, in the dialectic of the exchange of goods, the producer does, truly, *transcend* himself by being absorbed within a universal structure of production, mediated by the system of equivalencies that is imposed upon him in the objective form of a list of prices. It is no longer a matter, here, of the simple feeling of *honor*, in which the subject is still raised only to the immediate totality defined in terms of the actual field of exchanged goods: the abstraction of the *price* from then on permits him to dominate the universality of *possible* exchanges, and to absorb the collection of conceivable production in the ideal form of *justice*. In such a movement, each person takes on an absolute value as the authentic subject of a universal right, and the obligation of intersubjective help is laid down, not yet, of course, in the more profound form of charity – as love of the other *qua* other – but at least within the limits of the material conditions of existence that define the dignity of productive activity: "I have given food to those who were hungry, and clothing to those who were naked." Thus it is, indeed, in the process of economic life that ideal structures are engendered that give to human life its truly human meaning. But the truth which is strictly immanent to social reality, takes the form of a mystical transcendence. It is no longer a matter of this purely human sense of justice that is constituted, in a necessary manner, in the practice of commodity exchange, as at the preceding stages, where honor is constituted in the exchanges of goods, and duty in the exchanges of services. It is, on the contrary, the mystery of a *divine* justice that imposes its laws through threats and promises. In fact, the act of hoarding that is maintained under the movement of trade gives to this movement the negative form of a game of reciprocal concessions, and the real expansion of the individual, which is elevated to the universal structure of total production, is reversed in a simple, ideal restriction of his primitive egoism. Such a restriction can, evidently, be justified only on the plane of transcendence; more specifically, it is conceived only within the mystical horizon, from that time on, drawn by the privative form of appropriation, and the negation of hoarding operates within the general framework of hoarding itself. Thus, the insufficiency of production forces, which entails the maintenance of their sporadic form and thereby the objective contradictions from which exclusivistic behavior ensues, transforms the natural sense of justice that is engendered by the primordial movement of commercial relations into a supernatural order, which defines, in fact, only the equilibrium of real forces as the synthesis of practically possible hoardings. And it is the privative moment of such an appropriation that is exalted into a *divine justification* of which the privileged of this world try to assure themselves through all kinds of sumptuous monuments, rodomontade, and protestations of high virtue.

If we pass from Egypt to Mesopotamia, we again discover the same general structures, but affected by a particular characteristic. On the one hand, the religious syncretism sanctions here, as there, the political unification realized by absolute monarchy; on the other hand, the cult of the god who dies and rises from the dead, and of the goddess, his spouse and lover, confirms the consciousness of self, in its new mode of appropriation, through the symbolic recall of the alliance between the bourgeoisie and a fraction of the landed owners, which allows alliance for the victory over the ancient seignorial regime. Nevertheless, the corpus of these beliefs exhibits a vulgarity that contrasts violently with Egyptian moralism. The Sumero-Babylonian civilization overtly seeks the immediate enjoyment of goods, and its gods reveal a rather low, and even animal, nature. Thus, the tale of the Flood reports rather casually the panic that seized the gods when confronted by thunder and lightning: "The gods were afraid of the flood; they fled and ascended to the heavens of Anum. The gods crouched like dogs; they sleep outside." And when calm returned and after the hero who had escaped from the disaster offered a sacrifice, "the gods smelled the odor; the gods smelled the good odor; the gods assembled like flies above the sacrificer."

The Egyptian spiritualism reflected the absorption of bourgeois structures in the economic control of absolute monarchy, the real universalism of mercantile relations being interiorized, through divine mediation, in the pure form of morality. Such an equilibrium could be realized, evidently, solely owing to the natural protections that guarantee a relative isolation to the valley of the Nile. Yet Mesopotamia was open through a multitude of access routes that certainly favored a foreign invasion, in addition to commercial traffic – thus pushing class oppositions to extremes. On the one hand, the intensity of trade had, in fact, already produced in Mesopotamia the development of banking activities, while Egypt hardly went beyond the stage of the trade of commodities. On the other hand, feudal barbarism constantly reappeared in Mesopotamia with its original violence, in the form of invasion, and kept the bourgeoisie from constituting for itself a general conception of the world corresponding to the level of its real development. Indeed, while the exchange of commodities implies the still actual presence of natural determinations under the universal structures that are constituted upon them through trade practice, the movement of *credit* posits the universal in the pure abstraction of its intelligible being as an *in-itself* that is immediately available in financial relations. The development of economic life overflowed the framework of absolute monarchy, while absolute monarchy continued to be imposed by virtue of the importance of land property and of the barbarous environment: thus, the new structures were affirmed, in all of the brutality of their actual content, as practical structures of real activity, while the gods, on their part, retained their primitive vulgarity. Already, however, a new symbolism is outlined in which the appropriation of the forms of rationality transcends their juridico-economical content. *Destiny* is the concept of the universal imposing itself as pure external necessity, and the development of divination expresses the desire of appropriating this necessity for itself in the very form of its exteriority. But the actual appropriation supposes an interiorization in which the concept is affirmed *qua* concept. By taking hold of the power in the city, the Hellenic bourgeoisie will be raised to the universal *in and for itself*, absorbing the sensible figure of the divinity in its intelligible notion as the rationality of the *Logos*.

9. MONETARY ECONOMY, THE TRANSCENDENCE OF THE IDEA, AND THE CONCEPT OF SALVATION

If money (in its broadest sense, as a privileged commodity that serves as an intermediary) appeared as early as the stage of barbarism, its characteristic form as *coined money* did not become widespread until the first millennium, when trade developed on the plane of a universal economy. In the ancient Orient, payments were made with corn or metals that had to be measured or weighed each time, and if Mesopotamia already knew fiduciary money in the form of payable notes, its use remained, nevertheless, relatively limited. With the expansion of civilization that followed the great population movements of the second millennium, the intensity of trade induced the state to coin money in order to spare private individuals from constant weighings and measurings. Progress incurred its full development in the Hellenic cities – marking the decisive predominance of urban life over agricultural. Thereby, a new horizon was outlined in which the structures of universality were posited, at last, in the specificity of their meanings. This is the underlying reason for the 'Greek miracle': the Greeks, having attained civilization at a late date, profited from all that was acquired during the previous development, and the passage from mercantile to monetary economy was effected among them with an exceptional rapidity that allowed the bourgeoisie to liberate the rational forms that were characteristic of its mode of production from the ancient magico-religious horizon. *Money* is the movement of exchange that is reflected within itself in a disposable object – the mediation that raises the universality of goods that are still immediate and sensible to the autonomy of self-consciousness, *as a concept in the form of a concept.*

From the beginnings of urban development in Greece, class struggle occurs in a radically new guise. While in the Orient, the bourgeoisie had to reinforce monarchic power in order to insure the security of exchange against the seignorial nobility, in Greece it was the nobility itself that aided the trade movement and, in addition, being in favor of the power that the accumulation of movable wealth, along with landed property, gave it, overthrew kingship in favor of an aristocratic regime. In other words, profiting from the general expansion of international exchange and, furthermore, favored by the disposition of maritime routes, Greek trade was of sufficient account to attract the landed nobility, inducing it, thereby, to sacrifice its specifically feudal interests. Indeed, under these conditions, aristocratic power returned as a plutocracy, in which some members of the bourgeoisie of necessity participated, even if only through the interplay of matrimonial relations: "The man of good birth," complains Theognis, "does not refuse to take the daughter of a villain for his wife, as long as she brings him great wealth. By the same token, a woman consents to marry a villain, provided that he is rich. It is money that is esteemed; money mingles the races." Thus, by virtue of its historical position, Hellenic feudalism had favored the urban movement and opened the way for the bourgeois revolution. The stage of absolute monarchy was avoided, and the rapid progress of monetary economy weighted the relation of forces in favor of the bourgeoisie, which, from the sixth century, undertook the acquisition of political power. The scarcity of cultivated areas in Hellenic territories gave the urban structures an importance that they could not acquire in the great continental empires, and the ideological form of appropriation was freed from the mythical horizon characteristic of feudal structures, which was perpetuated in the Orient under the guise of monarchical power. The *divine being* takes the form of *intelligible being* – the eternal substance from which all particular existents are engendered and wherein the property of *money* is exalted as the universal mediator by the very efficacy of its mediating power.[1]

In Ionia, the movement occurs in its original authenticity. Against the resistance of a traditional, still powerful nobility, the revolutionary bourgeoisie of Miletus opposed the practical efficacy of its mode of production in the form of a materialistic monism – the diversity of sensible entities being resolved in the becoming of primordial matter, just as the diversity of concrete wealth is resolved in the becoming of *money*. The consciousness of a universal power that is directly available in the possession of money realized the divine in a positive form, the '*eternal and ageless*' substance, *athanatos kai ageros* – water, according to Thales, or the Boundless according to Anaximander, or indeterminate matter that Anaximenes designates as the air, that engenders everything through condensation and rarefaction.

Thus, bourgeois thought expressed its claim to absorb the particularism of landed property in the universalism of a monetary economy. The movement was reversed under Persian domination, since trade appeared to the nobility itself as the sole force that would allow the Hellenic city to resist foreign oppression effectively. Thus, the unity of money no longer had to be imposed upon the plurality of forms of wealth through the theory of a primordial substance, since accord was achieved in a subtler manner by the movement of plurality itself. Heraclitus of Ephesus, the die-hard aristocrat, dreamt of a sacred union in which members of the bourgeoisie and feudalists showed mutual respect for their respective interests in order to provide a common front against the invader: "Couples are things whole and things not whole, what is drawn together and what is drawn asunder, the harmonious and the discordant. The one is made up of all things, and all things issue from the one."[2]* Thus, universal mobility did not have for its purpose the dissolution of the real in the diversity of subjective appearances, but rather the rediscovery of true reality in the very movement of oppositions and contradictions; viz, the common interest of all classes in the oppressed city, such as is asserted obscurely throughout their particular conflicts. More specifically, it was a matter of exalting diversity as generative of unity, of saving what still remained of feudal particularisms in the name of the exigencies of the common struggle: "Men do not know how what is at variance agrees with itself. It is an attunement of opposite tensions, like that of the bow and the lyre. It is the opposite which is good for us. The hidden attunement is better than the open."[3]*

The philosophy of Heraclitus does not leave the plane of the real any more than does Milesian materialism, since the new structures preserve the oppression of their revolutionary meaning as a universality that asserts itself within the very materiality of practical life. At the other extremity of the Greek world, in the new cities of the Italian coast, the barbarian menace was hardly felt. On the other hand, since its population was composed of emigrants, feudal power had not been able to provide strong traditions for itself, and it was the bourgeoisie that, imposing itself upon the nobility, brought the nobility to its ideology for a common appropriation. Hence, the universal unity of monetary economy (which the revolutionary thought of Ionia posited within the horizon of the real, as primordial substance or unity immanent in the very flux of unseizable becoming) is no longer presented except in the form of a symbolic sanction of bourgeois aristocratic power. For the actual reality of the universal was denied and absorbed within a mystical transcendence – i.e., the *God of pure reason*, freed from all sensible attachment, in which is exalted the exclusivity of the act of appropriation that is accomplished in the very universalism of *monetary* relations: "One God," says Xenophanes, "the greatest among Gods and men, neither in form like unto mortals nor in thought .... He sees all over, thinks all over, and hears all over. But without fail he swayeth all things by the thought of him in mind."[4 *]

The conjunction of rationalism and mysticism in the thinkers of classical Greece marked the access of the bourgeoisie to the position of the ruling class in the form of an alliance with the nobility. The structures of intelligibility constituted in the new mode of production are detached from their authentic origin in order to be idealized in pure forms of justification, such as the supernatural properties of numbers of the Pythagoreans or the immutable One of the Eleatics, where the magico-religious significations characteristic of feudal appropriation are unexpectedly reborn. Thus, the liberation of rational thought that was attained in Ionian positivism turned out, with Parmenides, to be a spiritualism of a new kind, which absorbs the real movement of intelligibility; viz, the unifying movement as homogeneous with the multiplicity that it unifies in the transcendence of an ontological absolute: "Nor was it ever, nor will it be; for now *it is*, all at once, a continuous one. For what kind of origin for it wilt thou look for? In what way and from what source could it have drawn its increase? ... I shall not let thee say nor think that it came from what is not; for it can neither be thought nor uttered that anything is not. And, if it came from nothing, what need could have made it arise later rather than sooner? Therefore must it either be altogether or be not all."[5 *]

In such a return to mysticism, the revolutionary movement is reduced to a simple struggle of tendencies within the coalition of the ruling classes – the bourgeoisie imposing upon the nobility the form of unity that is characteristic of the process of money, but on a supernatural plane in which the negation of sensible becoming eliminates, in advance, the new levels that emerge in productive labor. Nevertheless, in the fifth century, the fact that monetary economy reached its full development opened the perspectives of a total rationality, and the democratic revolution was expressed in doctrines of pure immanence: the atomism of Leucippus and Democritus and the humanism of the Sophists. It was no longer a question of the bourgeoisie sharing power with the nobility, but rather of eliminating the nobility with the support of the common people, to the advantage of purely urban interests, and the form of transcendence that was inherited from the feudal era was annihilated in the face of properly human values created by civilization: "Man," says Protagoras, "is the measure of all things, of things that are that they are, and of things that are not that they are not." In fact, the power of money no longer consists in simply absorbing riches within the movement of trade; it outlines the framework of a new mode of existence that is founded on *technique* as it develops within the horizon of purely human experience. "There are among men," says Polos, "a host of arts drawn from experience by means of the experimental method: for experience directs life according to art, inexperience according to chance." From that time, revolutionary thought passed from speculations on substance to a concrete theory of knowledge and action; for the problem that was posed in reality was no longer to integrate the pluralism of landed property within the unity of monetary relations (that for which the alliance of the bourgeoisie with the nobility suffices, but rather to guarantee the total primacy of civilized forms, social law appearing as what alone distinguishes man from the animal:

> Now apply this analogy to our present condition [says Protagoras to Socrates in Plato's dialogue of the same name]. The man who in a civilized and human society appears to you as the most wicked must be thought just – a practitioner as one might say, of justice – if one has to judge him in comparison with men who have neither education nor courts of justice nor laws nor any constraint compelling them to be continuously heedful of virtue – savages in fact .... If you found yourself among such people ... you ... would bitterly regret the very depravity of our own society.[6 *]

Thus, in the movement of the monetary economy there is engendered a new world, in which mankind is elevated to the autonomy of a universal form. At the primordial stage of mercantile economy, the form of goods was still separated only from the concrete content of wealth, and *justice qua* law of exchange appeared as an extrinsic relation that could be guaranteed and interiorized only through the transcending mediation of divinity. With the use of money, every object is presented, from now on, as a commodity, since the universality of possible exchanges immediately posited it in the rationality of its *concept*. Hence, justice is no longer imposed as a divine command, but rather as a purely human truth that is constituted in the immanence of human relations.

As to the gods [concludes Protagoras], I can know neither whether they are nor whether they are not. Too many obstacles prevent a solution to the problem, such as the obscurity of the subject matter and the shortness of life.[7*]

Nevertheless, such a truth immediately turns into its opposite, and the critique of traditional values ends, in fact, in an inconsistent relativism – the new masters claiming the triumph of any thesis whatsoever, according to their whim. The humanism that was born within the democracies of the fifth century did, indeed, imply a profound contradiction in its actually real content. On the other hand, its universalistic intentions implied the suppression of social or national opposition in the expansion of civilized life. "By nature," says Antiphones, "we are in all respects identical at birth; we all breathe air through the mouth and nostrils." Thus, there is no reason whatsoever to "respect and honor people who are said to be of good birth and to despise men of low status." By the same token, "none of us has been originally distinguished either as a barbarian or Greek." But, on the contrary, the development of the monetary economy that inspired this universalism rested, in fact, on the industrial use of slaves. Yet slavery found its principal source in piracy and war: the slave is the barbarian; expressed more generally, the foreigner according to the original meaning of the term (i.e., the enemy); and his exploitation entailed the preservation of local and national particularisms along with the hierarchical traditions that support them. Thus, the actual conditions of ancient production did not enable the bourgeoisie to realize, within the framework of democracy, that universal order to which it aspired by virtue of the new structure of human relations that arose in the progress of the productive forces. The harsh labor of the slave was the *secret mover* that thrust the city, irresistibly, beyond its historical limits, while the citizen clung to the particularistic horizon of his mode of exploitation. Tragedy presented to consciousness the drama that rent consciousness in its actually real life – the destruction of traditional values symbolized by the misfortunes of the hero and interiorized in a pure self-delight in *terror* and *pity*. Thus, the cosmopolitanism of the Sophists did, indeed, correspond to authentic exigencies, but not yet to an actual form of their realization. It is for this reason that, in their practices, they were eager to suppress the sense of truth claimed in their theories, and man, as the 'measure of all things', was no longer man become universal through the work of civilization, but simply ambitious man who claims, through the subtlety of his arguments, "to make the weaker thesis appear the stronger."

With the flowering of Sophistry in Athens, Hellenic thought had exhausted its capacities for revolutionary development. The struggle of the bourgeoisie against feudalism was possible, to be sure, only through its alliance with the little people: yet, this alliance had meaning for it only insofar as it favored bourgeois commercial expansion, whose benefits largely covered the expenses incurred in the city by the practice of democracy. As soon as the conflict of the rival centers during the Peloponnesian war had defined the limits of such a system, the leading members of the bourgeoisie of Athens threw themselves behind the nobility and claimed to regain, through internal rebellion, what they had lost through external defeat: in the year 404, the oligarchic coalition took hold of the city under Spartan protection, celebrating its triumph on the ruins of the fatherland. However, the bloody excesses of the tyranny of the Thirty provoked the return of the democrats, and the oligarchy was eager to disavow the crimes that had been committed during its rise to power. Plato, who had participated in the civil war under the command of the Thirty tyrants, undertook to cover, under an idealistic cloak, the ambitions of the rebellion that failed – not failing, however, to search for the conditions that would enable its realization at a better time. The project of a perfect city, living in 'justice' and 'wisdom' had to be accomplished through the reciprocal concessions of the bourgeoisie and nobility, which would bring about their fusion into a new aristocracy. On the one hand, the bourgeoisie would renounce the economic development that had led to demoncracy:

> And men [says Socrates in the *Gorgias*] say it is these who have made made our city great, never realizing that it is swollen and festering through these statesmen of old. For they have paid no heed to discipline and justice, but have filled our city with harbors and dock yards and walls and revenues and similar rubbish, and so, when the crisis of her infirmity comes, they will hold their present advisors responsible and will sing the praises of Themistocles and Cimon and Pericles, who caused their misfortunes.[8]*

The nobility, for its part, must abandon the vulgarity of its traditional mysticism in favor of a purified spiritualism that is capable of absorbing already elaborated rational structures. In such a compromise, the conceptualism of the monetary economy lost its real meaning as the form of the expansion of production and exchanges, in order to be integrated within the transcendence of a pure *world of ideas* in which the universalism of bourgeois relations benefits, paradoxically, from the systematic exclusivism of feudal appropriation. In other words, the forms of intelligibility actually constituted in productive labor and the movement of exchanges are hoarded by an aristocracy of a new kind, which, holding the monopoly on the vision of the Ideas, will exert a limitless authority on the activity of the citizens. The concept of a government of philosophers is the ideal justification of the real movement in which the bourgeoisie returns to feudalism, and the suppression of the sensible datum in the ideality of Forms in themselves accomplishes the symbolic sacrifice that guarantees the privileges of the new nobility.

If we examine closely the actual content of the intelligible world, we will find there, to be sure, nothing more than what is already presented in the real world – the Idea representing, in a mystified form, the universal structure of goods to the extent that they are determined in kinds and species that are constituted as autonomous unities in the process of trade. The problems that arise in philosophical speculation, thus, concern actually real difficulties. More particularly, the discussions about the One and the Many reflect, very exactly, the relation of money to the plurality of commodities that it dominates. Thus the well known play of hypotheses of Plato's *Parmenides*

Table of the hypotheses of Plato's *Parmenides*

If the One is One		it is meaningless	(1st hyp.)
If the One is	*it*	is everything	(2nd hyp.)
	it	is nothing	(3rd hyp.)
	the Others	are everything	(4th hyp.)
	the Others	are nothing	(5th hyp.)
If the One is not	*it*	appears to be everything	(6th hyp.)
	it	appears to be nothing	(7th hyp.)
	the Others	appear to be everything	(8th hyp.)
	the Others	appear to be nothing	(9th hyp.)

merely describes with precision the various moments of a simple monetary economy. After a false start, in which it is demonstrated that money *qua* money taken in isolation (i.e., '*the One which is one*') could not be defined in any way and thus, strictly speaking, is devoid of meaning, the discussion moves on to the consideration of actual relations – the movement of trade in which money functions as the unifying unity (i.e., 'the One which is'). And the conclusion is reached that whatever the hypothesis, whether the *One* is or whether it is not, we will obtain a symmetrical play of contradictory consequences – as much for the *One* as for *Others*: in other words, as much for money as for commodities. To be sure, the reality of money, once posited (i.e., 'if the One is'), will obviously assume all possible determinations within the dialectic of exchanges in which its being consists precisely in being identified with the diversity of commodities: it will, therefore, be unity and multiplicity, identity and difference, etc. (2nd hypothesis). If, however, we take the very *instant* of the passage in which the act of exchange consists, money will be neither money nor commodity – no longer being the one and not yet being the other: it will, therefore have no determination whatsoever and will be nothing (3rd hypothesis). If we now consider the form of the commodity, we see immediately that it bears all intelligible determinations – i.e., the same as money, since it is exchanged for it (4th hypothesis); but, on the other hand, it does not bear any intelligibility, since the commodity is a sensible reality that exists in itself, outside of commercial relations (5th hypothesis). Remarkably, if we suppress the reality of money (i.e., 'if the One is not'), the same consequences return, at least on the plane of appearance: for, the exchange of commodities anterior to the constitution of coined money, which defines the money, already produces the same structures, even though money alone can give the characteristic of being – i.e., of availability. Thus *Parmenides* concludes: "whether there is or is not the *One*, both that one and the *Others* alike are and are not, and appear and do not appear to be, all manner of things in all manner of ways, with respect to themselves and to one another."[9]*

Such an impasse betrays, on the philosophical plane, the insufficiency of the mediations acquired in real development. The movement of trade consists, to be sure, in a constant exchange between unity and multiplicity, identity and difference, but it does not allow the mastery of the very foundation of the relation as *exchange value*, or, in philosophical language: the nature of the One *at the instant of the passage*. We indeed know that the exchange value of an object is defined in terms of the quantity of social labor that is necessary for its production. But in the older economy, productive activity being confused with the status of the producer *qua* slave, serf, or artisan in a corporation, could not be the object of a rational calculation, and the appropriation of the movement by the privileged classes reduces it to the pure, abstract transcendence of the model in which the universal form of the command is defined – the Idea that imposes upon the real the meaning of its being. Thus, the reason for the equivalences remained hidden, and the relation of the One to the Many appeared, in the free game of trade, as a simple given whose discursive intelligence remains powerless to recover its foundation. Such a system could function as long as the problem of the *content* was being covered up by the prosperity of exchanges. Indeed, during the period of the expansion of the city, trade developed, at the expense of the barbaric world, that guaranteed both large profits and the furnishing of servile man-power, and a simple abstract legality was sufficient to maintain the equilibrium of relations between citizens. But liberal democracies rapidly found their limitations in the very logic of their development – the exploitation of new countries, introducing civilization in them and placing them in a position that rivalled the ancient centers. The diminution of benefits that resulted thereby posed the problem of an organization of the actual *content* of exchanges – in other words, the problem of the passage from unity to multiplicity, for which, as we have seen, there was no rational solution in the ancient world because the structure of productive labor lay concealed under the still immediate form of exploitation, and exchange value presented itself as a mystery: the *One* which, at the instant of the passage, excludes every determination. Hence, the Platonic notion of Justice – no longer as the abstract form of exchanges but as the concrete unity of social functions – can be defined only in terms of the pure ineffable transcendence of the Idea of the *Good*, the intuitive knowledge of which enables the philosopher to claim limitless power in the city.

To the extent that the problems of organization will find their empirical solution in a regression toward monarchic absolutism, finality will descend from the Idea into the real and the world of sense perception – i.e., the world unified by the authority of the monarch, which will be identified by Stoicism with the very being of Reason. The synthesis of bourgeois rationalism and feudal mysticism that, in Plato, still respected the specificity of the concept as the unifying unity, ends up in a systematic confusion with the passive notion of a providential *Destiny* that justifies the citizen's resignation in the face of royal power.

Thus, in the decadence of the city the bourgeoisie clung to its privileged positions by returning to more and more elementary forms of transcendence. The movement ends with Christianity, where, under the pressure of Roman imperialism, Hellenic thought made a supreme effort to save its rational values on the very plane of irrationality. As pure intuition, *Charity* expressed that universal unity of content that was outlined within the horizon of monetary economy without finding the necessary mediations to be realized effectively. Thus, the real movement of appropriation must be proposed as the sanction of its privative form – the exclusivistic ideal of a Kingdom of Heaven that was reserved only for the *elect*. And because the death of God (which did not pose any problem for traditional beliefs, accustomed as they were to attribute human adventures to the divinity) became incomprehensible after the purging of Hellenic thought, the concept of *incarnation* succeeds in giving a coherent aspect to the divine sacrifice, as the symbol of universal sacrifice that guarantees to the bourgeois, in the very universalism of civilized forms, the security of hoarding.

The remarkable development of Mediterranean trade, from which Christianity garnered the spiritual heritage, was but a particular aspect of the general expansion of exchanges that spread throughout the whole of Asia and during the first millenium awakened India and China to monetary economy. In these last two countries, just as in Greece, the specifically revolutionary movement expressed itself in sophistry, while the actual establishment of bourgeois property by the monarchical power found its ideological reflection in the doctrines of Salvation. In India, the warring nobility, itself oppressed by the theocracy of the Brahmans, participated in the struggle of the bourgeoisie against the rigid distinctions of the caste system and thereby restricted the import of the universalist movement to a purely negative form of deliverance – the theoretical suppression of the principle of individuation in Buddhist mediation. In China, the absence of a caste system permitted the class struggle to develop in a more normal way. Undoubtedly, the real perspectives of the bourgeois revolution did not yet go beyond the monarchical stage, which implied the preservation of the feudal horizon. Nevertheless, within this very framework, the struggle against the particularism of the lords permitted the new form of appropriation to be affirmed directly in an ideal appropriation of *universal Life* as the mystical symbol of the universal power of money.

Thus, throughout the diversity of local doctrines, the universal concept of Salvation expressed the universality actually acquired in human relations through the expansion of monetary economy. It is rather obvious, though, that whatever the importance of the feudal concepts preserved in it as the reflection of admitted concessions by the bourgeoisie to the nobility for the establishment of the new order under the absolute authority of the monarch and, also, whatever the role played by the progressive elements of the nobility, the great forms of spiritualism that, in the history of self-consciousness, marked the passage from barbarism to civilization must be related to bourgeois thought, itself the expression in this era of the general interests of mankind. However, the authentic meaning of human universality appears in it only under the form of its negation: more precisely, the real suppression of immediate determinations by productive labor and the movement of exchanges is systematically confused with the ideal suppression of objective reality with the symbolism of sacrifice. *Renunciation* is the mystical mediation that transforms the real universality of human relations elaborated by civilization into a purely ideal eternity where the enjoyment of universal values is found to be reserved, paradoxically, for a small number of *saints* and *elect*. In such a mystification, the doctrine of Salvation betrays itself as a justification of the privative form of appropriation on the very plane of universality. The ideal hoarding of spiritual life by the saints and elect is but the *solemn complement* of the real hoarding of economic goods by the exploiting classes. Indeed, the juridical relations, as the recognized form of real hoarding, imply the spiritualization of the object that is no longer the object that is, in fact, usable by all, but the *sacred* object (i.e., the object rendered untouchable by the defense that is pledged by the proprietor). But if, at the level of feudal economy, the diversity of wealth still implied for each productive unity only a private defense symbolized by sacrifice to the protecting God, the constitution of the universal value of money required for the security of its appropriation an agreement of all on a universal defense – in other words, a general *renunciation* that guarantees to everyone the exclusive possession of his own goods. In its daily form, such a renunciation still assures only the good functioning of contractual right: in other words, the hoarding of the social resources of monetary economy by the coalition of feudalists and bourgeoisie under the aegis of absolute monarchy. In its exalted form, it becomes the *pure mystical renunciation of the goods of this world*, in which the ideal hoarding of these very same goods takes place, transfigured by the pure symbolism of an *eternal life*. And, in fact, such an eternity is reserved, precisely and in all exclusivity, for the very same ones among whom the movement of appropriation has been frustrated to the point of accepting all the sacrifices, and this eternity, to be sure, produces this very transfiguration. Yet the exalted form is the complement of the everyday form, since it gives it the prestige that enables it to be perpetuated. Thus, the pure mysticism of the great religions of Salvation is, to be sure, a necessary moment within the dialectic of privative appropriation – the spiritualistic model of hoarding that maintains social unity by calling on the exploited to secure for himself the conditions of his own exploitation in the hope of a total appropriation by means of the virtues of renunciation.

10. CAPITALISTIC ECONOMY, THE POWER OF ABSTRACTION AND THE PROLETARIAN REVOLUTION

The Christian Middle Ages were the actual realization of Plato's *Republic*, with its double aristocracy of clerks and warriors dominating the mass of workers. The return to feudal barbarism was achieved within the universalistic horizon elaborated by the ancient bourgeoisie and maintained by the power of the Church. Such a *hierarchy* of the internal structures defined the general framework that would permit the Western bourgeoisie to pass on to the capitalistic stage. Indeed, owing to the ecclesiastical organization and the relative unity that it assured, the rebirth of trade could profit from the very beginning from the coming into existence of the whole European market, even if to bypass it; while production, by virtue of its feudal assignment, was still being organized only within the particularistic framework of the commune. The disproportion between the volume of international demand and the scarcity of resources of local handicraft (at least for certain special products such as drapery) induced the merchant to intervene more and more in the very process of productive labor by furnishing the raw materials and sometimes the tools themselves and by giving the orders throughout the various stages of the manufacturing process. Thus, in Flanders and Italy, starting with the thirteenth century, there were constituted the first capitalistic structures that reduced the craftsman to simple wage-laborer and absorbed his activity in the abstract rationality of an *economic calculus*. However, it was still but a question of sporadic formations that threatened the established relations without being able to be constituted within a new horizon.

But the intensity of the exchanges within the framework of Christianity had also led to a general progress of the productive forces, which, toward the end of the fifteenth century, placed the Western bourgeoisie in a position to corner the world market. The fabulous wealth obtained through piracy and merciless exploitation – sometimes even through the extermination of entire populations – provoked, at once, a considerable increase in the demand for manufactured objects, thus giving a decisive impulse to capitalistic production. Then, there was explicated the new form of bourgeois appropriation in which the universal power of money consists no longer simply in the unifying unity that extends to the plurality of commodities but in a *capacity of rational development* that is articulated precisely *qua* capital in the different moments of the productive process and, that allows the universal power of money to be mastered within an intelligible form. In such a movement, productive labor is revealed as the essence of value – the *universal constituent* that founds the unity of the multiplicity within the dialectic of exchanges. However, *revelation* is accomplished once more, only in the mode of dissimulation; for this labor is itself taken only in its pure hoarded form as that which generates *profit*, and the process is detached from its actual reality in order to be reduced to a pure mathematical necessity as exclusive of all human signification. In this way there is constituted the mechanistic horizon in which the modern science of nature will develop. If we consider the authentic meaning of scientific thought, it is rather evident that the foundation for its truth can be found only in *social human labor* – the ideal movement of the concepts of mathematical physics being but the abstract symbol of the real movement of industry as the *practically realized* agreement of the intellect and things. But this same labor is divested of its human quality in the appropriation by capital, and the abstraction covering up the dialectic that founds it (viz, scientific truth) assumes the paradoxical appearance of a negation of lived experience:

> In production [says Marx], intellectual powers expand in one direction because they vanish in many others. What is lost by the detail laborers is concentrated over against them in capital. It is a result of the division of labor in manufacture that the laborer is brought face to face with the intellectual powers of the material process of production, as the property of another and as a ruling power. This process of separation begins in simple cooperation in which the capitalist represents to the individual laborer the unity and the will of the social body of labor. It is developed in manufacturer that deforms the laborer into a detail laborer. It is completed in large industry, which makes science a productive force distinct from labor and presses it into the service of capital.[1 *]

Thus, the power of abstraction is substituted for the traditional forms of transcendence as the justification of the privative movement of appropriation; and within the pure operative necessity of the mathematical concept the new bourgeoisie exalts the magical transmutation of the creative power of human labor into a simple *commodity*, as *pure matter* capable of being exploited in the *economic calculus*. During the whole course of the preceding development, the appropriation of labor by the ruling class enveloped the very person of the laborer in a spiritualistic horizon that enabled everyone to *recognize* himself in a world full of meaing, and the movement of mystification maintained the intimate relations of consciousness with its object in the very form of negation and transcendence. With the constitution of capital there is accomplished the division between the laborer *qua* pure subject of right and his objective content – the labor force that he has sold to the capitalist and the struggle against the statutory relations of domination to servitude generalizes the mechanism of the economic calculus in the conception of a world *destitute of meaning*.

However, the practical conditions of an actual security forced the bourgeoisie into an alliance with the former exploiting classes; and, if the mechanistic abstraction was entirely valid for the exploitation of the proletariat, the coalition of the exploiters could guarantee the solidarity of its various factions only by restoring in some manner the spiritualism of traditional forms. During its first revolutionary period, bourgeois thought, being elaborated within the framework of absolute monarchy, claimed to defend, with the existence of God and the immortality of the soul, the general structure of medieval mysticism while at the same time giving it a new signification with mathematical rationalism, in which the new mode of exploitation is justified. The pure interiority of the *cogito* preserved, with the exclusivity of privative appropriation, the *subjectivity* of the possessors against this very *objectivity* with which they claimed to dispose of the real producers. And when, on the point of taking hold of political power, the French bourgeoisie exalted the mechanism of capital in the concept of the object as pure matter, Kant, rendered more prudent by the backward status of Germany, posed the problem of the *conditions of possibility* of such an object – in other words, the problem of a *juridical guarantee* of bourgeois appropriation. It is rather obvious that if, on the plane of pure right, the abstract form of universality was given its own foundation by means of the autonomy of its ideal movement (i.e., the symbolic work of the bourgeoisie in the economic calculus), the conditions of an actually real assurance imposed an agreement with the aristocracy by a restoration of the theological concepts under the double disguise of a 'subjective finality' and a 'practical postulate'.

In any case, however, the movement of abstraction was imposed as long as the general interests of mankind were confused with those of the bourgeoisie in its revolutionary struggle against the incoherences of feudal status. But the victory of the bourgeois revolution immediately posed the problem of the *actual content of universality* – in other words, the problem of a practical organization of the new legality. The Hegelian dialectic represented, in the passionate idealism of a *universal reconciliation*, the rallying cry of the former exploiting strata to the bourgeois order, in which they hoped to find the full satisfaction of their economic interests, precisely by renouncing the politico-religious pretensions in which the particularism of their traditional forms of appropriation was extolled. Such an adhesion *qua suppression and self-realization* is presented as the supreme term of a spiritual becoming whose ideal articulations reflect the successive revolutions in which the former ruling strata resigned each time by the very reason of their material interests, in order to be absorbed in a more efficacious mode of exploitation – in other words, in a *more elevated spiritual form*. The phenomenological dialectic is the living calling to mind of the historic movement of the struggles and coalitions that enabled the bourgeoisie to surmount, progressively, the statutory power of the feudal chiefs, narrowly tied to the contingency of natural determinations, in order to be imposed fully with the pure abstract universality of the capitalist system – an abstraction that, nevertheless, is covered over with the paradoxical appearances of a *concrete Universal* precisely by virtue of recollection:

The goal, which is Absolute Knowledge or Spirit knowing itself as Spirit, finds its pathway in the recollection of spiritual forms as they are in themselves and as they accomplish the organization of their spiritual kingdom. Their conservation, looked at from the side of their free existence appearing in the form of contingency, is *History*; looked at from the side of their intellectually comprehended organization, it is the *Science* of the ways in which (phenomenological) knowledge appears. Both together, or History (intellectually) comprehended, form at once the recollection and the Golgotha of Absolute Spirit, the reality, the truth, the certainty of its throne, without which it were lifeless, solitary, and alone. Only

The Chalice of this realm of spirits
Foams forth to God His own Infinitude.[2*]

Thus, the triumphing bourgeoisie claimed to *preserve* the mysticism of the previous forms of oppression in the very movement of their *suppression*, symbolizing thereby the cooperation that the traditional formations were to bear for the common security of private property. However, the objective structure of capitalistic exploitation dissolved every combination of this kind through its very own foundation; for the feudal prestige that the bourgeoisie would have wanted for its own had been made possible only through the subtle hierarchy of complex statutes that capitalism annihilated precisely through the cold rationalism of relations of money:

The bourgeoisie, wherever it has got the upper hand, has put an end to all feudal, patriarchal, idyllic relations. It has pitilessly torn asunder the motley feudal ties that bound man to his 'natural superiors', and has left no other bond between man and man than naked self-interest, than callous 'cash payment'. It has drowned the most heavenly ecstasies of religious fervor, of chivalrous enthusiasm, of philistine sentimentalism, in the icy water of egotistical calculation. It has resolved personal worth into exchange value, and in place of the numberless indefeasible chartered freedoms, has set up that single, unconscionable freedom – Free Trade. In one word, for exploitation veiled by religious and political illusions it has substituted naked, shameless, direct, brutal exploitation.[3*]

Such is the actual content that the *dialectic of real history* unveiled in the triumph of the spirit celebrated within the idealistic dialectic; and the whole movement of traditional spiritualism appears, henceforth, in the practical experience of the proletariat as so many forms of the alienation of human values authentically elaborated in productive labor. Since the origins of mankind with the constitution of the first production techniques, we have seen being engendered at every level a real form of universality in which the individual actually *transcends* himself in the process of cooperation, such as is required by the exigencies of production itself. In this way is defined the true foundation of spiritual structures that gave to human life its properly human meaning. *Duty* is the simple form of unity of collective labor. *Honor* defines the immediate collaboration between independent productive unities. *Justice*, as the mediated structure of cooperation between distinct producers, is born from the use of money, which elevates the productive relations to a rational structure. However, such a determination still solely concerns the abstract form of the movement: the organization of production in the actuality of its human content defines the meaning of *charity* where the universality of mediation rests on the immediate being of the individual taken as a *social* producer. But the real level of the productive forces maintained them in a state of dispersion that implied, ineluctably, with the opposition of the various productive unities, the privative form of appropriation exalted in the symbolism of transcendence, and the actual meaning of universality, constituted in the experience of cooperation, was transposed into pure mythical functions: totemism and the cult of spirits for the still immediate spirituality of duty and honor, the submission to the all-powerful God, and the imitation of the Savior God for the mediated universalism of justice and charity. However, such an alienation in which the practice of exploitation is sanctioned, preserved (notwithstanding) some semblance of authenticity on the subjective plane; for the movement of universality could still be affirmed only within the very framework of private property that continued to be asserted by virtue of the separation of the productive centers.

With the expansion of capitalism and large industry, there is elaborated a new structure of productive activity, which implies the bringing of the economic organization into a unified whole. Within the horizon of the proletariat, which experiences such conditions as the very truth of existence, the meaning of universality is constituted in a radically new form that no longer is presented to the singular egos as the ideal movement of their suppression, but implies their actually real suppression through the suppression of the privative form of appropriation itself. Private property (which was, since the birth of mankind, with various accommodations, the permanent condition of the development of productive forces) became in itself an intolerable hindrance; and the movement of transcendence (which was perpetuated from the time of the *mana* of the primitives to the purified concepts of the philosophers as the universal foundation of all signification and value) betrays itself, henceforth, as the pure symbolism of hoarding, which justifies the desperate resistance of the propertied classes to the establishment of the new humanism. The materiality of productive labor appeared to the property owner as the very form of his 'alienation', and the whole effort of his spiritual life strove to surmount such an 'exteriority' in order to 'return to himself' in an idealism that guaranteed the security of a purely exclusive enjoyment. But, if such an exclusivity produced the form of exploitation, everyone – exploiters and exploited – was unanimous in the common aspiration for private property; and the conflicts that arose from the mode of appropriation did not, in any way whatsoever, affect its privative form. But the progress of the productive forces raised them to the level of a total universality that, hoarded in the process of capital, definitively takes away from the real producer any hope of appropriating for himself on the individual plane the result of his production. It is then that the exclusivism of privative appropriation is revealed as dupery and *alienation*; and the *return to the self* no longer consists in inventing a new form of transcendence that enables the hoarding movement to be adapted to the universality of social labor, but rather, in guaranteeing an *appropriation without exclusion*, with the recognition of the material reality of productive activity as the foundaton of its truth-meaning. More precisely, the meaning of property is confused, henceforth, with the very labor of production within the horizon of human universality, and dialectical materialism expresses the movement of human values that are engendered in social human labor. Hence the merciless critique that it makes of the concepts of the former society is no longer a simple destruction, but rather a restitution of their authentic content, which enables the new humanity to take them up again on a superior plane. In the construction of socialism and the passage to communism is realized, finally, that *universal reconciliation* which was the dream of bourgeois thought in the idealistic dialectic of the forms of exploitation and which the proletariat places on its true ground by means of the organization of social labor, where every class structure and every pretext of exclusivity is suppressed. As the realization of the human form of humanity, Marxism achieves the ideal aspirations of the past. But this itself is not asserted in terms of an idea: it is the actual movement of the social totality where the traditional formations are absorbed in the proletariat by the *materiality* of their real life.

APPENDIX

Table of sensori-motor development

Phylogenesis	Protista	Sponges	Coelenterates	Worms	Fish	Mammals	Lower Apes	Anthropoids	Man	
Ontogenesis			0	1 month	5 months	9 months	1 year	18 months	2 years	
Behavior	Attraction and Repulsion	Contraction	Reflex Displacement	Locomotion	Apprehension	Detour Manipulation	Intermediary	Instrument	Tool	Language (predicative)
Lived Meanings		*Impression*	*Sensation*	*Sensorial Field*	*Object-Phantom*	*Real Object (Thing)*	*Real Relation*	*Image*	*Representation*	*Concept*

NOTES

AUTHOR'S PREFACE

[1] The publication of the manuscripts is being prepared at the Husserl Archives in Louvain under the direction of Professor H. L. Van Breda.
[2] *Ideen*, p. 319. [*Ideas*, translated by Boyce Gibson (New York, 1962), p. 92.]
[3] *Ibid.*

PART ONE

1. THE TECHNIQUE OF VARIATION

[1] *Logische Untersuchungen,* Volume I, Third Study, Chapter 1 (1901).
[2] *Ibid.*, Volume II, Third Study, Chapter 1 (1901).
[3] *Ibid.*, No. 6.
[4] We understand the act of constituting as a *theme* of study. The impossibility of separating color from extension is *experienced* in the lived consciousness of self. Empiricism comes to a realization of this impossibility in psychological concepts and speaks of the impossibility of abstracting two sensations – or rather two aspects of the same sensation. The phenomenologist thematizes it in intentional concepts and discloses the *meaning* of that lived-consciousness of impossibility: it is color itself which is inseparable from extension.
[5] The most important texts are the following: *Ideen*, No. 70; *Formale und Transzendentale Logik*, No. 98; *Cartesian Meditations*, No. 34; *Erfahrung und Urteil*, No. 87 to 98.
[6] *Erfahrung und Urteil*, p. 412.
[7] We must distinguish between 'the thing' as material object (*das Ding*) and 'the thing itself' (*die Sache Selbst*): this latter case concerns every object in general inasmuch as it can be 'given' in the corresponding intuition. Husserlian terminology leads to confusion in regard to the term '*Sache*', using it alone, as designating the simple 'thing', to the exclusion of its predicates of value and action.

2. PURE IDEALITIES AND EMPIRICAL IDEALITIES

[1] Cf. Paul Natorp: 'Husserls "Ideen zu einer reinen Phänomenologie" ', *Logos* 7, p. 228 ff.
[2] *Formale und Transzendentale Logik*, No. 63 and 87; *Erfahrung und Urteil*, p. 397. On the active character of the *Wesensschau*, cf. Fink, 'Die phänomenologische Philosophie in der gegenwartigen Kritik', *Kantstudien* 38, p. 329.

[3] The term *ideality* must be used in preference to *universality*: the purely intelligible content of a singular object is defined as an 'essence of substrate' (*Substratwesen*) – *Ideen*, No. 14.
[4] *Erfahrung und Urteil*, pp. 396–397.

3. THE TRUE SIGNIFICANCE OF THE NOTION OF ESSENCE

[1] 'Philosophie als Strenge Wissenschaft', *Logos* 1, pp. 303–313; *Ideen*, No. 8 *et seq.*
[2] *Ideen*, No. 153; *Cartesian Meditations*, No. 59.
[3] We should note that, from a strictly logical point of view only an analytic necessity can be found on the plane of pure possibility. Variation can make precise the content of a concept, only by eliminating all relation to existence. If we wish to understand the *synthetic a priori* value of the laws of essence, we then must return to the *origin of the possible*, in the experience of *the real itself* inasmuch as it discloses itself in the meaning of its being. The imagination operates on schemata which, themselves, imply an anterior perception, with the structure of the being itself defining the form in which we can imagine it.

4. DIFFICULTIES WITH THE OBJECTIVISM OF ESSENCES

[1] 'Worldly' refers to everything that is anterior to the transcendental reduction, which is the precise means of our getting out of 'the world'. Material and formal aprioristic sciences are related to the world inasmuch as the essences that they study are still essences *of* this world (*Formal and Transcendental Logic*, pp. 198–199).
[2] The total system of all the forms of possible theories constitutes the *Mannigfaltigkeitslehre*. Such a project is revealed as incapable of realization, with the works of Gödel. Nevertheless, this idea remains valuable for our meaning from the philosophical point of view. The definition of *Mannigfaltigkeitslehre* makes explicit only the ideal signification of mathematical activity inasmuch as it describes sets of objects – 'multiplicities' – through the law of operations that one can perform upon them. A method such as this, once having reached its 'saturation point', has to be able to allow us to determine in an unequivocal manner all the individual members of the set: every proposition having a meaning in the system will be capable of being demonstrated to be true or false. The impossibility of arriving at a saturation point for every system which is equal or superior to arithmetic, means simply that the ideal cannot be realized; nevertheless, it retains its value as such. The opposition between logicism and intuitionism seems to us to retain a confusion of the philosophical plane (in which 'regulative ideas' must be defined), with the mathematical level (in which only those operations that can be realized are taken into account). Cf. *Logische Untersuchungen*, Vol. 1, No. 60; *Ideen*, No. 72; *Formale und Transzendentale Logik*, Nos. 28 to 36. On Gödel's theorem, see Jean Cavailles, *Axiomatic Method and Formalism*, pp. 144–151.
[3] Plato: *'Parménide'*, 130 c and d. We know how Plato, through the notion of a diallectical genesis, tried to reintroduce multiplicity and becoming into his system. Division allowed the Idea to give itself a content by its own movement. Nevertheless, this process remains on the plane of the purely intelligible and always lacks relation to existence.

The exigency of such a relation led to the contradictory idea within Platonism of a rehabilitation of the sensible. The *Philebus* had to appeal to pleasure in constituting the Sovereign Good. [Plato: *The Collected Dialogues*, ed. by E. Hamilton and H. Cairns, Bollingen Series, Pantheon Books; cf. 'Parmenides', 130 c and d, p. 924.]

5. THE RETURN TO LIVED EXPERIENCE

1 Husserl had received his doctorate in 1883, having written his thesis on mathematics: *Contribution to the Theory of a Calculus of Variations*. [*Beiträge zur Variationsrechnung*]
2 *Logische Untersuchungen*, Vol. 1, No. 71.
3 The logico-mathematical notions received from that time a subjective definition; i.e., entailment consists simply in our gathering together in thought certain contents – in our thinking them in a single act (*Philosophie der Arithmetic*, p. 79).
4 *Logische Untersuchungen* (1900), Vol. 1, Preface to the First Edition, p. vii.
5 *Die Krisis der Europaischen Wissenschaften und die Transzendentale Phänomenologie*, III A, p. 156 of the transcription, note. It is concerned with the sequel to the article published in *Philosophia* 1.
6 *Ibid*. It must be understood that, in the phenomenological reduction, human subjectivity is 'reduced' to absolute transcendental subjectivity.
7 *Logische Untersuchungen* (1901), Vol. 2, p. 4.
8 *Ibid*., p. 7.
9 *Ibid*., p. 9.
10 This table that I see is a sensible object. But the fact that it is rectangular is a 'categorial object', which can be grasped only in predicating it: 'This table is rectangular'. In sensible perception we have the real table with its form, but not the intelligible 'state of affairs' (*Sachverhalt*) expressed in the judgment.
11 *Logische Untersuchungen*, Vol. 2, Sixth Investigation, No. 45 ff. The notion of 'foundation' brings to mind the Gestaltist relation of background to 'figure'. The categorial object is detached, so to speak, from a sensible background. But it is suitable to specify the relation as a relation of knowledge: its predication expresses the *truth* of immediate experience.
12 The proposition is true for all cases, including that of the laws of formal logic. Valid for an object in general, such propositions refer still, in a sense, to the real, since the object is always an *object of the world*. For this reason, they are not, despite their purity, absolutely independent of sensible content: in Husserlian language, 'their kernels are relevant' (*Formal and Transcendental Logic*, No. 87).
13 *Logische Untersuchungen* (1901), Vol. 2, p. 18.
14 *Ibid*., p. 21.
15 '*Reell*', which we translate as '*real*' (in quotation marks), designates the actual content of lived experience, in opposition to the object intended ('intentional content'). In the Husserlian vocabulary, it is opposed strongly to natural reality (*real*).

6. THE DISCOVERY OF THE REDUCTION

1 The text has just been published in the second volume of *Husserliana*, under the title *Die Idee der Phänomenologie*. [1951-Ed.]

[2] *Die Idee der Phänomenologie*, p. 31. [*The Idea of Phenomenology*, translated by W. Alston and G. Nakhnikian (The Hague: Martinus Nijhoff, 1964), p. 24.]
[3] *Ibid.*
[4] The Interpretation of the Cartesian *cogito* as a universal *cogito* surely, does not make any sense whatsoever. It cannot be a matter of *my* existence, actually lived: "*I am, I exist*; that is certain. But for how long? As long as I think; for it could be the case that if I totally ceased to think, I would, at the same time, cease to be altogether" (*Second Meditation*).
[5] *Die Idee der Phänomenologie*, p. 44.
[6] 'Pure' simply means 'non-worldly'. Contrary to Kantian usage, the concept designates a concrete content.
[7] *Die Idee der Phänomenologie*, p. 44.
[8] *Critique du Jugement*, Introduction, No. 4. [*Critique of Judgment*, translated by James Meredith, (Oxford, 1964), p. 19.]
[9] The text, which dates from 1909, has the notation: F_1 17.
[10] More precisely, the '*sum res cogitans*'; in '*cogito ergo sum*', the *sum* designates no other than the existence of thought in its actually being lived, or, in Husserlian terms, 'transcendental existence'.
[11] F_1 17, pp. 34–38 of the transcript.
[12] *Die Idee der Phänomenologie*, pp. 56–57. [*The Idea of Phenomenology*, pp. 44–45.]
[13] *Ideen*, No. 60.

7. THE EXPOSITION OF THE *IDEEN*

[1] *Ideen*, p. 49. [*Ideas*, translated by Boyce Gibson (New York, 1962), p. 92.]
[2] *Ibid.*
[3] *Ibid.*, pp. 54–55. [*Ibid.*, p. 98.]
[4] *Ibid.*, p. 59.
[5] *Ibid.*, p. 59. [*Ibid.*, p. 98.]
[6] *Ibid.*, Nos. 37–38.
[7] *Ibid.*, pp. 75–76. [*Ibid.*, p. 119.]
[8] *Ibid.*, p. 86. [*Ibid.*, p. 131.]
[9] *Ibid.*, pp. 91–92. [*Ibid.*, p. 137.]
[10] *Ibid.*, p. 59.
[11] *Ibid.*, pp. 83–84. [*Ibid.*, p. 128.]
[12] *Méditations Cartésiennes*, Strasser edition, p. 75; Levinas' translation, p. 31: "As an Ego, in the natural attitude, I am likewise and at all times a transcendental Ego, but I become aware of this only in effecting the phenomenological reduction." [*Cartesian Meditations*, translated by Dorion Cairns (The Hague, 1960), p. 37.]
[13] *Ideen*, p. 92.
[14] *Ibid.*, p. 60.
[15] *Nachwort zu meiner 'Ideen zu einer reinen Phänomenologie und phänomenologischen Philosophie'*, p. 8. [*Ibid.*, p. 9.]
[16] Sensible data are 'profiles' or silhouettes (*Abschattungen*) of a sort which the external object casts on us.

[17] *Ideen*, No. 55. [*Ibid.*, p. 153.]
[18] "It then becomes evident to us that Kant's intuition rested on this (phenomenological) domain, although he had not yet appropriated it and had not yet succeeded in recognizing it as the ground for the study of a science of autonomous and rigorous essences. Thus, the Transcendental Deduction of the first edition of the *Critique of Pure Reason*, for example, moves (properly speaking) on phenomenological ground, but Kant mistakenly interpreted it as psychological and permitted it to escape notice once again. It evidently concerns the well-known passage of the three syntheses."
[19] *Ibid.*, p. 302. [*Ibid.*, p. 166.]
[20] Landgrebe: 'Husserls Phänomenologie und die Motive ihrer Umbildung', *Revue Internationale de Philosophie* **2**, pp. 300–301.
[21] Cf. the remarkable story by Hering, '*La Phénoménologie d'Edmond Husserl il y a trente ans*', *ibid.*, p. 366 ff.

8. THE CRITIQUE OF THE KANTIANS

[1] Natorp: 'Husserls Ideen zu einer reinen Phänomenologie', *Die Geisteswissenschaften* **1** (1914); *Logos* 7 (1917–1918).
[2] In opposition to the objectivation or construction of the object in the knowledge of nature, the *subjectification* or constitution of the subject will be defined as a *reconstruction*. The latter is the task of psychology, such as Natorp understands it, and he relates it to the phenomenology of the *Ideen*. This inspiration is found in Kant's famous text of the three syntheses (*Allgemeine Psychologie nach Kritischer Methode*, Chap. 8).
[3] Rickert: 'Die Methode der Philosophie und das Unmittelbare', *Logos* **12**, p. 235; and 'Vom Anfang der Philosophie', *Logos* **14**, p. 121. We know that the defense of the critical point of view is, for this author, accompanied by a remarkable effort to give a positive reply to the reproach of formalism: the notion of *value* allowed for the restitution of a 'content' to philosophical reflection.
[4] Kreis: *Phänomenologie und Kritizismus* (1930); Zocher: *Husserls Phänomenologie und Schuppes Logik* (1932).

9. FINK'S REPLY

[1] Eugen Fink: 'Die phänomenologische Philosophie E. Husserls in der gegenwartigen Kritik', *Kantstudien* **33** (1933), p. 319. The article is preceded by a preface in which Husserl indicates total approval for the work of his assistant.
[2] Rickert's school had defined the *a priori* as *meaning*: its idealism is defined as an 'idealism of meaning'.
[3] In fact, the use of the term in the *Ideen* refers expressly to the theory of knowledge (*Ideen*, p. 59).
[4] Fink, *op. cit.*, p. 341.
[5] The theory of knowledge searches for the conditions which ground the possibility of the object: it concerns the *essence* which is known to be still in the world.
[6] *Ideen*, p. 319.

10. THE NOTION OF CONSTITUTION

[1] *Méditations Cartésiennes*, translated by Levinas (1931), p. 78. Strasser edition (1950), p. 126. [*Op. cit*., p. 94.]
[2] *Ideen*, Vol. 2, Third Part, Appendix VII.
[3] *Ibid.*

11. THE CONSTITUTION OF THE WORLD OF THE SPIRIT

[1] We recall that the concept, in the phenomenological sense of the term, refers to the conditions of evidence: a 'founded' existent is an existent which can be 'perceived' only on the 'foundation' of another existent which is, for this reason, qualified as 'founding'. Thus the perception of the soul of another takes place on the 'foundation' of his body. The other modes of knowledge (e.g., memory or judgment) refer to that original mode, as condition of a possible verification.
[2] *Ideen*, Vol. 2, Third Part, No. 1.
[3] *Ibid*., First Part, Nos. 2–7.
[4] *Ibid*., Vol. 2, Third Part, No. 2.
[5] *Ibid*., No. 3.
[6] *Ibid*., No. 6.
[7] *Ibid*., No. 13.
[8] *Ibid*., Appendix VII.

12. THE NOTION OF OBJECT

[1] We neglect spatial differences which introduce an important complication without modifying the foundation of things: the temporality of the most far-removed bodies refers to the perception of a possible consciousness, which, in the final analysis, has meaning only in my actual consciousness.
[2] The world of perception, or *life-world* (*Lebenswelt*) necessarily implies a *subjectivity* and a *relativity* which do not suppress its *reality* but define its *meaning*. It is the ever present universe of real things, given in the *experience of life*, which forms the ground (*Boden*) upon which the constructions of science are erected. The negation of the naive intuition in scientific truth, necessarily starts from this very intuition, and it is the lived reference to that ever living foundation which enables truth to be posited as truth of *this world*. Thus, it would be false to refer the 'subjectivity' of the *life-world* to 'psychology': the psychological is of the 'objective' and presupposes the constitution of the world as conceptualized by science; whereas it is a matter here of what is given to us *before* scientific objectivity. There would no longer be any point in speaking of a 'rhapsody of sensations'. The *life-world* presents itself as a horizon of objects, which have their own proper structure, a spatio-temporality, a substantiality, and a causality which are not yet scientific but which, nevertheless, imply an actual coherence and unity. It is this moving and relative unity which characterizes the life we all live and upon which the idealizations of theoretical thought are irreducibly dependent. The doctrine of the *Lebenswelt* as ground and foundation of all ideal activity is presented in the *Krisis*, notably in the first half of the unpublished part (*Krisis*, III, A). The con-

crete descriptions are found in the manuscripts of groups C, D, and E – i.e., constitution of the givens of immanence, of the primordial world, and of the experience of the other.
3 *Logische Untersuchungen*, Vol. 2, Sixth Investigation, Chapter 6.
4 *Erfahrung und Urteil*, p. 313. Cf. *Méditations Cartésiennes, op. cit.* "The supratemporality of ideal objects is demonstrated as *omnitemporality*, as a correlate of free producibility and reproducibility at all times."
5* Kant: *Critique of Pure Reason*, translated by N. K. Smith (New York, 1961), p. 124.
6* *Ibid.*, p. 135.
7* *Ibid.*, cf. p. 170, footnote.
8* *Ibid.*, p. 171, footnote.
9* *Ibid.*
10* *Ibid.*, p. 218.
11 The phenomenology of the *life-world* will take precisely the name of 'transcendental aesthetics' (*Formale und transzendentale Logik*, p. 256).
12* *Critique of Pure Reason*, p. 169, footnote.
13* *Ibid.*, p. 168.
14* *Ibid.*, p. 169.
15* *Ibid.*, p. 378, footnote.
16* *Ibid.*
17* *Ibid.*, p. 168.
18 *La Religion dans les limites de la simple raison*, First Part; General Observation. [*Religion Within the Limits of Reason Alone* (New York, 1960), pp. 42–43.]

13. SELF-EVIDENCE AND TRUTH

1 *Ideen*, p. 300 (English edition, p. 369).
2 *Formale und Transzendentale Logik*, p. 276.

14. THE PROBLEM OF ERROR

1 *Ibid.*, Nos. 105–106.
2 *Ibid.*, No. 58, p. 140.
3 The conception of the *Weltkonstitution* as *transcendental genesis* has its origin in the investigations of 'genetic logic', the subject matter of many unpublished manuscripts of 1919, summarized by Landgrebe in *Erfahrung und Urteil*. The general theory of genesis is outlined in *Formale und Transzendentale Logik* (1929) and *Cartesian Meditations*, written during the same period. (Cf. the 4th Meditation.)
4 R. Descartes: 'Meditations on First Philosophy', in *The Philosophical Works of Descartes*, tr. by E. Haldane and G. R. T. Ross, The Fourth Meditation (Dover, 1931), p. 175.
5 The major theory of the *Living Present* (*Lebendige Gegenwart*), hardly mentioned in the published works, is the object of particularly important and minute developments of Group C of the unpublished works. The 'living present' is the movement of primordial consciousness which is not *in time* but *constitutes time*: it is the *temporalizing temporality* always present to itself and which always is renewed in an eternity of life. While

* [Translators' note.]

consciousness which is *in time*, is itself constantly lost, the temporalizing modes of the *Living Present* represent a preservation and a perpetual conquest of self: the past is retained therein as that which *still is* (retention) and the future is announced therein as that which *already is* (protention). In this continual movement in which each present moment immediately passes into retention and sinks more and more into the past, but into a past which *still is*; while the future, *here and now* possessed in protention is actualized in a lived present – the self remains identical to itself, while renewing itself constantly; it remains precisely the same only by always becoming another, in that *absolute flux of an eternal Present*. Thus, primordial temporality is not a succession *in time*, since the past is *always there* and the future *already there* in the totality of an ever present actuality; but *time is* constituted in it as a succession of distinct moments, for the past is preserved only by virtue of being past, as that which *is no longer*, and the future is announced only by virtue of being future, as what *is not yet*. Thus, 'eternal' consciousness could not be separated from the consciousness which is lost in time. It preserves with it the relation of the *temporalizing* to the *temporalized*; seeing that time is constituted in a movement that is *itself temporal*.

Thus, the 'transcendental consciousness' described in the *Ideen* is not absolute consciousness. It is still in *time* – albeit in the purely 'phenomenological' time of lived immanence, which no clock could possibly measure. It is solely in this 'immanent' time that 'transcendent' time of external things is constituted. But transcendental immanence, *constitutive* in relation to the natural world, is itself *constituted* in the absolute subjectivity of 'time-consciousness', or 'the consciousness which constitutes time', in the primordial flux of the *Living Present*. This is, evidently, the major reason for the lack of comprehension which the text of the *Ideen* encountered, even though Husserl took care to specify that he voluntarily placed himself within an abstract point of view. (Cf. *Ideen*, p. 163.) "The transcendental 'Absolute', which we have thematized through the reductions, is not, in truth, ultimate; it is something which in a certain profound and wholly unique sense constitutes itself, and has its primeval source in what is ultimately and truly absolute." (*Ibid*., pp. 245–246.) "Any single lived experience which we may select shapes itself within a continuous 'primordial' time-consciousness as unity stretched out in phenomenological time. But we have expressly excluded this primordial synthesis of the original time-consciousness, along with the forms of inquiry which belong to it. The syntheses which we wish to speak of now are not those found within this *time-consciousness*, but within the limits of *time itself*, of concretely filled phenomenological time; or, as amounts to the same thing, the syntheses of the lived experiences – plain and simple – taken as we have so far always taken them, as unities having duration, as perishing processes in the stream of experience, which is itself none other than the filled phenomenological time."

The relation of constituting and constituted *time-consciousness* and *time* itself within phenomenological immanence, was described for the first time in the *Lectures* of 1905 on the *Phenomenology of Internal Time-Consciousness*, edited by Heidegger in 1928. In this study, Husserl takes a particularly 'noematic' standpoint, defining phenomenological *time* with its lived past and future, in terms of its 'appearing modes' (*Erscheinungsweisen*) in time-consciousness. The lived experiencing which endures in immanent time is an 'immanent object' which appears in profiles (*sich abschattet*) within the absolute flux of time-consciousness – its past phases being 'still present' in retention and its 'coming' (*das Kommende*) phases 'already present' in protention.

But the 'noetic' standpoint barely entered upon, immediately indicates a mystery: time-consciousness as the movement of retentions, of acts of consciousness of the 'now' and of protentions can be defined only as a '*flux*' which is not, nevertheless, a succession 'in time'. It is the absolute subjectivity whose very properties cannot be named: *Für all das fehlen uns die Namen* (No. 36).

The theory of the *Living Present* gives a first solution to this mystery.

"All in one (i.e., the absolute in its unity): unity of an absolute temporalization of an autotemporalization – the absolute is temporalized in its temporal modes of past, present, and future; is found in the absolute stream of the primordial Present, of the 'living and moving' Present (*strömend lebendigen Gegenwart*), of the Present of the absolute in its unity, which temporalizes within itself and has temporalized everything that is." (C_1 (1934), p. 4 of the manuscript.)

"My living and moving Present, my Present in its primordial mode, bears within itself all conceivable being: it is the primordially temporal temporality that is supratemporal (*die urzeitliche 'überzeitliche Zeitlichkeit'*) which bears within itself all times and all content in time as fixed temporal order (*alle Zeit als verharrend-seiende Zeitordnung*). Is it possible that I might never have begun to exist? But does the notion of a 'beginning' have any meaning if it is not that of a beginning *in* time? I, undoubtedly, can have a 'first awakening' and a 'last awakening'. But can I have a beginning as 'Living' Present? Without this Living Present, nothing would have existence – neither other 'selves' which are implied in it, nor the world with the birth and death of man in it." (C_2, III (1932), pp. 8–9.)

"Thus, we must make a distinction: (on the one hand) we have the temporal stream of consciousness, related to this temporality, the transcendental self of lived acts, of faculties, of habits, which has constituted a spatio-temporal world, and always continues to do so; (on the other hand) we have as the primordial foundation of this temporalization and of all 'transcendent' temporalizations (as opposed to 'immanent' temporalizations), the primordial self, grasped concretely as the self of this life which temporalizes all existence (*als Ich dieses all-zeitigendes Lebens*) . . . . The concrete flow of the Present is revealed as the primordial phenomenon; transcendental subjectivity in the primordial form of its being is the Present as primordial stream (*urströmende Gegenwart*) in the form of the being of a flow which flows (*in der Seinsform des Strömendverströmens*) but as Present and always and anew Present is constituted constantly. As we shall see, 'Present' has a figurative sense. For its proper being, as transcendental being in the primordial form, is not in the ordinary sense a present as a determined temporal moment flowing between a past and a future which flows with it. However, the use of this word is unavoidable. In order to characterize the concept, we use the expression *primordial*, concrete Present [as primordial (urphänomenale)]. Every transcendental problem resulting from the phenomenological reduction refers, in fact, to this 'primordial phenomenon'. In it arise perceptions; even better, it itself is, in a sense, a *perception* in its totality and in every moment which constitutes its being, in the upsurge of its lived experiences and of their moments. For, that which belongs in an actually 'real' (*reell*) manner to this total sphere is conscious and precisely conscious in the primordial mode, and this is equally the case for it as a totality . . . .

"Being as primordial phenomenon, as living and moving Present, is primordially conscious – a field of primordial perception with all its components. This present has an astonishing structure which it is appropriate to follow in various directions and, first of all, in that which characterizes it as a flow. Its profound essence is to be constituted as the *nunc stans* of a unique flow through an anonymous continuity of intentional modifications of a primordial mode – a continuity which is not fixed but which itself flows. And in this flow is constituted a *primordial Now which is posited* as permanent (*stehendes und bleibendes Ur-Jetzt*), as fixed form for a content which flows, and as the starting point of all constituted modifications. With the fixed form of the primordial now is constituted a double continuity of equally fixed forms; in this manner there is constituted on the whole a fixed continuity of the form (of time) – a continuity in which the primordial now is the focal point from which two continuities as branches of the flowing modes are derived: the continuity of moments which have just passed and that of moments which are coming. This is a continuous form which is maintained permanently and through which flows a content which is always constituted with it. It is this flow through the form of time that a surprising synthesis constantly unfolds, in which every individual being is constituted, arising primordially in the now, and always remaining identical with it by passing through the system of forms of 'what has just passed' in the modes of past (as always progressively more past). [The Living Present is the ever present actuality of an absolute movement through the continuity of retentions and protentions which flow, as intentional modifications that are themselves present from the primordial (i.e., present) mode. In this movement which is not in time, is constituted time as 'the *a priori* form of inner sense' with the continuity of the forms of the *now* and of the *past* 'which has just passed' and of the *future* 'which is coming'. It is in this immanent form which is already *constituted* that the unity of the transcendent temporal object, in its turn, is constituted, because of a particular synthesis which maintains it as 'the same' through the successive moments of objective duration.]

"The Present which flows [i.e., Living Present] is the Present of the movement of flowing, of having flowed, and of having yet to flow (*Gegenwart des Verströmens, des Abströmens und des Zuströmens*). The now, the continuity of the past, and the living horizon of the future, which is outlined in protention, are conscious 'at the same time', and this 'at the same time' is an 'at the same time' which flows (*dieses Zugleich ist strömendes Zugleich*) . . . . The 'Living Present which flows' is not the stream of consciousness (*'lebendig strömende Gegenwart' ist nicht Bewusstseinstrom*).

"This primordial movement is not, absolutely speaking, in any time, for time finds its source only in it." [C_2, I (1932–1933), pp. 12, 13, 15, 16, 21, 22.]

It is hardly worth noting the *dialectical* character of these descriptions. The movement of the *Living Present* consists in repeatedly *going beyond itself* by retaining within itself its past in terms of its *annulled, preserved*, and *elevated* moment. The Hegelian concept of *Aufhebung* acquires all of its concrete value exclusively from the primordial stream which is annulled at each instant and by this very annulment is realized in its true being. It is this movement as 'pure restlessness of life and process of absolute distinction," which *constitutes time* in its different modes: the *past* as what is left behind, the *present* as the very leaving behind, inasmuch as it is *actually* accomplished, the *future* as what *comes to* consciousness inasmuch as it transcends itself *toward* itself. In retention, the past is maintained *in itself*; in protention, the future is announced *for itself*. The transcendence of the past toward the future constitutes the actuality of the present, as *the passage of the in-itself to the for-itself*. By becoming *for itself* that which it was *in itself*, consciousness is realized in its truth. The *being of the object* is precisely nothing but this very truth inasmuch as it is revealed *to consciousness*: it is the *being for consciousness of its in itself*. Thus, the world is constituted in an absolute movement of which time, as a succession of evanescent moments, is but the abstract appearance: "The phenomenon is the process of arising into being and passing away again, a process that itself does not arise and does not pass away, but is *per se*, and constitutes reality and the life-movement of truth. The Truth is thus the Bacchanalian revel where not a member is sober, and because every member no sooner gets detached than it *eo ipso* collapses straightaway, the revel is just as much a state of transparent unbroken calm." (Hegel: *Phénomenologie de l'Espirit*, Préface.) [*The Phenomenology of Mind*, translated by J. B. Baillie (New York, 1966), p. 105.]

15. [SELF-]EVIDENCE AS INTENTIONAL PERFORMANCE

1 *Ideen*, Nos. 99–117.

2 We obviously give only very general outlines here with the aim of specifying the meaning of these notions. Real cases are more complex, presupposing the whole *sedimented* acquisition of previous developments. Hence, one could object that it is sometimes jealousy which gives rise to love. To which we would answer that a phenomenon of this kind always presupposes a previous love, constituted in an unauthentic manner – indeed, purely symbolic – and which jealousy has come to awaken and renew. We can be jealous only in regard to a being upon whom we think we have claims. Such a conviction can be founded on an actual love. It can be implied simply in our relations with this person. But, inasmuch as we consider the person as a being who belongs to us, at least in principle, we view him or her with a meaning which carries with it an implicit act of love and constitutes it from then onwards as lovable. It is only upon the foundation of this meaning that a situation of rivalry can arise with the answer that it contains. Love is made explicit and developed in the form of a '*reactivation*'; we become conscious of 'always' having loved this person – the circumstances alone allowing us to reveal ourselves to ourselves.

3 *Logische Untersuchungen*, Vol. 2, Sixth Investigation, No. 25. [English edition, *op. cit.*, p. 739.]

4 Cf. A, V_5, p. 10 of the manuscript: "The self is always a pole; it has no extension, nothing which implies an exteriority of *partes extra partes*; it is not a whole formed by pieces connected to each other: it is possible, nevertheless, to make an analysis of its egological being, but it is not a construction of parts that would have been placed in it as if it were done in space. Its description leads ineluctably to the temporality of consciousness and to the movement by which it temporalizes itself as a quasi-extension of time upon the whole of its egological life – its taking a position, its position in regard to values . . . . The being of the self is not, as the non-self, permanence through changes but permanence in the preservation of self, and to preserve oneself is to conserve oneself with constancy and to acquire oneself by perpetuating what is acquired in the self, by preserving it in itself."

5 *Formale und Transzendentale Logik*, p. 250.

6 *Méditations Cartésiennes*, Nos. 37–38.

16. THE POSSIBILITY OF ERROR

1 Phenomenology, with the notion of intentionality, revives the theory of association. It is not a matter of a composition of sensorial givens, whose truth and objectivity are

incomprehensible; rather, it is a matter of the *transfer of a meaning*, previously constituted on a given configuration, to a similar configuration. But the meaning is itself the result of a *primordial instituting* (*Urstiftung*) in (self-)evidence. "Every apperception in which we apprehend at a glance, and noticingly grasp, objects given beforehand (e.g., the already given everyday world), every apperception in which we understand their sense and its horizons forthwith, points back to a '*primordial instituting*' in which an object with a similar sense becomes constituted for the first time. Even the physical things of this world that are unknown to us are, to speak generally, known in respect of their type. We have already seen like things before, though not precisely this thing here. Thus, each everyday experience involves an analogising transfer of an originally instituted objective sense to a new case, with its anticipative apprehension of the object as having a similar meaning. To the extent that there is giveness beforehand, there is such a transfer. At the same time, that sense-component in further experience, which proves to be actually new, may function in turn as institutive and found a pre-giveness that has a richer meaning. The child who already sees physical 'things' (*Dinge*) understands, let us say, for the first time the meaning-usage of a pair of scissors; and from this time he sees a pair of scissors at the first glance *as* a pair of scissors, but naturally not in an explicit reproducing, comparing, or inferring" (*Méditations Cartésiennes*, No. 50). We recall that 'thing' (*Dinge*) must be understood to be the sensible material object, whose unity, proper to naive perception, is not yet due to the forms of intelligibility of the judgment. This objective unity, immediately given to intuition itself, refers to a *genesis*, in which it is constituted as arising from sensorial givens and sensible configurations, which do not yet possess the meaning of the reality of the 'thing'. "With good reason it is said that in infancy we had to learn to see 'things' (*Dinge*), and that, on the genetic plane, such an experiment must precede all other modes of being conscious of 'things'. For, the field of perception, such as we find it given in our infancy, does not yet contain anything which could be perceived, at first glance, as 'thing'. (*Méditations Cartésiennes*, No. 38.)

[2] Cf. Kant: *Critique of Pure Reason*, Preface to the second edition: "In the earliest times to which the history of human reason extends, mathematics, among that wonderful people, the Greeks, had already entered upon the sure path of science, but it must not be supposed that it was as easy for mathematics as it was for logic – in which reason has to deal with itself alone – to light upon, or rather to construct for itself, that royal road. On the contrary, I believe that it long remained, especially among the Egyptians, in the groping stage, and that the transformation must have been due to a revolution brought about by the happy thought of a single man, the experiment which he devised marking out the path upon which the science must enter, and by following which, secure progress throughout all time and in endless expansion is infallibly secured. The history of this intellectual revolution – far more important than the discovery of the passage round the celebrated Cape of Good Hope – and of its fortunate author, has not been preserved. But the fact that Diogenes Laertius, in handing down an account of these matters, names the reputed author of even the least important among the geometrical demonstrations, even of those which, for ordinary consciousness, stand in need of no such proof, does at least show that the memory of the revolution brought about by the first glimpse of this new path, must have seemed to mathematicians of such outstanding importance as to cause it to survive the tide of oblivion. A new light flashed upon the mind of the first man (be he Thales or some other) who demonstrated

the properties of the isosceles triangle. The true method, so he found, was not to inspect what he discerned either in the figure, or in the bare concept of it, and from this, as it were, to read off it properties; but to bring out what was necessarily implied in the concepts that he had himself formed *a priori*, and had put into the figure in the construction by which he presented it to himself. If he is to know anything with *a priori* certainty he must not ascribe to the figure anything save what necessarily follows from what he has himself set into it in accordance with his concept." (*Critique of Pure Reason*, English edition, translated by Norman Kemp Smith, p. 19.)

3 Benedict Spinoza: *Ethics*, Part II, Prop. 35, Note.

17. A DIGRESSION

1 René Descartes: 'Meditations on First Philosophy', in *The Philosophical Works of Descartes*, translated by E. Haldane and G. R. T. Ross (Dover, 1931), the 'Third Meditation', Volume 1, p. 158.

2 *Ibid.*, 'Fifth Set of Objections', Volume 2, p. 151.

3* *Ibid.*, 'Reply to Objections V', Volume 2, p. 213.

4 *Ibid.*, the 'Third Meditation', Volume 1, pp. 158–159.

5 *Ibid.*, the 'Fifth Meditation', Volume 1, p. 184.

6 *Ibid.*, the 'Second Meditation', Volume 1, p. 150.

18. PHENOMENOLOGICAL DESCRIPTION

1 *Ideen*, No. 24, *Das Prinzip aller Prinzipien.*

2 *Formale und transzendentale Logik*, p. 12. [*Formal and Transcendental Logic*, translated by Dorian Cairns (The Hague: Martinus Nijhoff, 1969). This translation has indicated page references to the German original in parentheses.]

3 *Ibid.*, pp. 3, 4, 5, 14.

4 *Ibid.*, p. 217.

5 *Ibid.*, p. 218.

6 *Ibid.*, p. 280.

7 It concerns the synthesis by association of passively engendered meanings with the present 'givens'.

8 *Méditations Cartésiennes*, No. 38. (*Cartesian Meditations*, translated by Dorian Cairns (The Hague: Martinus Nijhoff, 1960), pp. 70–80.)

9 *Formale und Transzendentale Logik*, pp. 277–278. (Cairns' translation, *op. cit.*, p. 316.)

10 *Méditations Cartésiennes*, No. 18, Strasser edition, *Husserliana* 1, p. 81: "*Das universale cogitatum ist das universale Leben selbst . . . .*" (Cairns' translation, *op. cit.*, p. 43.)

19. THE CONSTITUTION OF THE FORMAL DOMAIN

1 *Fomale und Transzendentale Logik*, No. 25. For more details, cf. *Erfahrung und Urteil*, No. 61 and 96, a.

[2] We are using 'predicative' in its broad sense, considering the operations of conjunction, numeration, etc. as entering partially into 'predication', properly speaking. (*Formale und Transzendentale Logik*, p. 97.)
[3] *Formale und Transzendentale Logik*, p. 88.
[4] *Ibid.*, Nos. 42–43.
[5] *Ibid.*, Nos. 44–45. *Apophansis* is to be understood exactly as the intended meaning as such, as the noematic content of the act of judging.
[6] *Ibid.*, Nos. 52 and 54.
[7] *Formale und Transzendentale Logik*, No. 87.

20. THE GENESIS OF JUDGMENT

[1] This work that was published in 1939 has the subtitle: *Untersuchungen zur Genealogie der Logik.*
[2] *Erfahrung und Urteil*, p. 132.
[3] *Ibid.*, pp. 242–243.
[4] *Ibid.*, pp. 242–246.
[5] *Ibid.*, p. 277.
[6*] *Critique of Pure Reason*, p. 167.
[7] *Erfahrung und Urteil*, p. 313.
[8] We can usefully compare this description with Hegel's dialectic of consciousness, where the same movement is presented on the conceptual plane. Cf. the introduction to *Phénoménologie de l'esprit*. "This *dialectic* process which consciousness executes on itself – on its knowledge as well as on its object – in the sense that out of it the new and true object arises, is precisely what is termed *Experience* . . . . Consciousness knows *something*; this something is the essence of what is *per se*. This object, however, is also the *per se*, the inherent reality, *for consciousness*. Hence arises the ambiguity of this truth. Consciousness, as we see, has now two objects: one is the first *per se*, the second is the existence *for consciousness* of this *per se*. [The first *per se* is the immediate object of sensible activity: *the being for consciousness of this per se* is the meaning *passively preconstituted* in this activity in which the *per se* is revealed to consciousness in its true being.] The last object appears at first sight to be merely the reflection of consciousness into itself, i.e., an idea not of an object, but solely of its knowledge of that first object [i.e., the passively preconstituted meaning is *experienced* in consciousness as the *immanent meaning* of its activity upon the object: consciousness is conscious of it in a 'reflexion' on itself]. But as was already indicated by that very process, the first object is altered; it ceases to be what is *per se*, and becomes consciously something which is *per se* only *for consciousness* [i.e., the immediate object of sensible activity has taken on the meaning passively preconstituted in it. It thus exists no longer as objective reality that is *per se in that very meaning*, by which it is revealed to consciousness in its true being: it is the *per se for consciousness only*]. Consequently, what this real *per se is for consciousness* is truth: which, however, means that this is the *essential* reality, or the *object* which consciousness has [i.e., consciousness then returns to the preconstituted meaning and *constitutes* it actively by positing it as the very meaning of the object. What consciousness now perceives as objective reality is no longer the immediate object of sensible certainty – the first *per se* – but this object in the sense

that it has meaning for it as a subject which has predicates: '*the existence for consciousness of that per se*'. It is this logical being which, henceforth, will define the *true* or the *essence*]. This new object contains the nothingness of the first; the new object is the *experience* concerning that first object [i.e., with the passage to predication, the immediate world of sensible certainty has been suppressed: it is *experienced* in the course of sensible activity as *true being* such as the judgment, henceforth defines it. And so from now on the intuitive givens are perceived solely in the intelligible unity of the concept: in Kantian language, 'the unity which the object makes necessary can be nothing other than the formal unity of consciousness in the synthesis of the manifold of representation'].

Cf. the respective English translations of the above quotations. Kant's *Critique of Pure Reason*, p. 135, and Hegel's *Phenomenology of Mind*, pp. 142–143.

9 Hegel: *Phénoménologie de l'esprit*, from the Introduction. [*Phenomenology of Mind*, p. 144.]

10 If in the syntax of judgment we distinguish the form from the matter, the latter will be constituted with 'formations of the kernel' which will again consist of matter and form. Thus the proposition '*all bodies are divisible*' implies as *syntactic forms* (on the one hand), the forms of subject and predicate and their relation and as *syntactic matter* (on the other hand), the formations of the kernel, '*bodies*' and '*divisible*'. The kernel formation 'bodies' implies in its turn a form of the kernel (*the substantive form*) and a matter of the kernel (the bodies given in the antepredicative intuition). Thus the forms of the kernel applied to the antepredicative objects make them available for predication: they realize the Kantian definition of the *categories*: "they are concepts of an object in general by means of which the intuition of an object is regarded as determined in respect to one of the logical functions of judgment." (Cf. Kant, p. 128).

11 *Critique of Pure Reason*, p. 159.

21. THE GENESIS OF ANTEPREDICATIVE EXPERIENCE

1 *Méditations Cartésiennes*, No. 61.

2 Cf. 'Die Welt der lebendigen Gegenwart', in *Philosophy and Phenomenological Research* 6 (1946), 3.

3 Cf. above, Section 16, Note 1.

4 'Ursprung der Geometrie', in *Revue Internationale de Philosophie* 2 (January, 1939), pp. 216 and 223–224. [English translation by David Carr in *The Crisis of European Sciences and Transcendental Phenomenology* (Evanston: Northwestern University Press, 1970), pp. 366 and 375–376.]

5 *Erfahrung und Urteil*, p. 42 and 44.

6 *Formale undTranszendentale Logik*, p. 245.

7 'Grundlegende Untersuchungen zum phänomenologischen Ursprung der Raumlichkeit der Natur', in *Philosophical Essays in Memory of Edmund Husserl*, p. 324. The manuscript is of 1934 and summarizes: "*Umsturz der Kopernikanischen Lehre* in der gewöhnlichen weltanschaulichen Interpretation. Die Ur-Arche Erde bewegt sich nicht."

8 *Ibid*., p. 323.

9 Friedrich Engels: *Ludwig Feuerbach et la fin de la philosophie classique allemande*, Vol. 2 (Paris: Éditions sociales), p. 17. It should be noted that existentialism, by positing

an *essential* difference between human existence and natural reality, returns, in fact, to the plane of naive consciousness, namely the plane of *being* that is abstractly considered in its different modes. Thus it fully deserves the reproach of *psychologism* that Husserl makes in the well known text of the *Nachwort*. The Heideggerian notions of transcendence, situation, choice, project, etc., are *anterior to the reduction* and only systematize in an Aristotelian fashion the commonsense point of view. They presuppose this separation of domains that defines precisely the naiveté of the dogmatic attitude. On the contrary, dialectical materialism takes the different forms of being in the movement by which they pass from one to the other. Matter is posited in its *becoming conscious*, nature in its *becoming human*. Thus we find ourselves on a plane that is *posterior to the reduction*, the latter having suppressed the abstract conception of nature but not of actually real nature whose development implies the whole *movement of subjectivity*.

PART TWO

1. CONSCIOUSNESS AND MATTER

[1] Leibniz: *New Essays on Human Understanding*, contained in *Leibniz, Selections*, edited by P. P. Wiener (New York: Charles Scribner's, 1951), p. 391.

[2] *Lettre du 15 mai 1643.*

[3] *Lettre du 17 juin 1643.*

[4] *Critique of Pure Reason*, Norman Kemp Smith translation cited above, p. 474.

[5] The chemical phenomenon of *autocatalysis* already actualizes the characteristic structure of biological phenomena: thus a sprinkle of pepsin added to pure pepsinogen transforms the original into pure pepsin. Or one could very well express this by saying that the molecule of pepsin is multiplied at the expense of pepsinogen nourishment. [See the work of Prenant.]

[6] A goodly number of scholars are opposed to the heredity of acquired characteristics by reason of the finalistic background that it seems to imply. In fact, the process is perfectly rational, since it concerns only [as Lysenko tried to make clear] new characteristics *that introduce modifications into the formation of cells by the interchange of matter*. Everyone knows that mutilations in parents are not transmitted to their children. But an adaptation of the individual organism can provoke, at the level of the tissues that it concerns, particular secretions that, in their turn, bring about a modification of the corresponding factors in the reproductive cells. A particular role in this process would belong to the acids of the nuclei: we know that certain bacteria (i.e., colibacilli and pneumococci) pass from the rugose form to a smooth form when they are placed in the presence of cadavers of the smooth form or of an extract of these. The active principle that is responsible for this phenomenon of 'induction', is constituted by a highly polymerized thymonucleic acid that is produced by the bacterial cadavers in question (Avery-Boivin). It is conceivable that comparable processes are taking place in reproductive cells – these being essentially constituted of large nucleoproteic molecules similar to the protein viruses.

[7]* Engels: *Dialectics of Nature*, edited and translated by C. Dutt (New York: International Publishers, 1940), pp. 182–183.

3. THE MOVEMENT OF THE INTERNAL SENSE

[1] The capacity to reproduce former reactions is an absolutely general biological property as the experiences of habit formation in the infusoria have shown. Leucocytes of frogs have even been trained to react to light (Sgonina, 1938). Here we do not put to use the period of 'conditioning' that we prefer to reserve for more properly subjective phenomena.

[2] To be sure, sensible qualities are defined as modes of attraction and repulsion, which become specified within the dialectic of behavior. Sight, which begins to be sketched out in the medusa, refers back to the act of displacement passing to locomotion. Hearing and olfaction develop in fish as the results of locomotion being absorbed in apprehension.

4. THE MOVEMENT OF THE EXTERNAL SENSE

[1] Of course, we are here taking into account only whole levels of behavior. In fact, the utilization of tools does appear sporadically in lower apes. Weinert cites the case of Capuchin monkeys that make use of a stone to crack nuts.

[2] Köhler tells us, "As Sultan was much more expert than the others, in the beginning, he was often obliged to be present without helping, as I wished to ascertain the capabilities of the others .... If the observer's vigilance is at all relaxed, and the veto on building not continuously renewed, Sultan does not venture to enter fully into the work, but he cannot keep from 'lending a hand' here and there, supporting a box that threatens to fall under some adventurous and decisive effort of another animal, or otherwise taking a less important part in the work .... When we watch him, squatting beside the other animal, following all Grande's movements with his eyes and often with slight sketchy movements of arm and hand, there can be no doubt that these proceedings *in themselves interest him* and to a very high degree; that he follows and 'feels' the movements himself, and all the more keenly as they grow more difficult and crucial. The 'help' he offers at the critical moment is simply a *heightening of his already indicated participation in the process*: and interest in the other animal can play only a very secondary part, for Sultan is a pronounced egoist .... We are all acquainted with similar states of mind. It is difficult for anyone who, as a result of long practice understands any form of work, to stand aside while another bungles it: his fingers itch to intervene and 'do the job'. And we human beings, too, are far from wishing to help such a bungler from motives of pure altruism (our feelings towards him at the moment are not particularly cordial). Neither do we seek some external advantage for ourselves: *the work* attracts and dominates us." (Köhler: *L'intelligence des singes supérieurs* (Paris: Alcan), pp. 158–159.) [English original, quoted above: *The Mentality of Apes* (London: Routledge and Kegan Paul, Ltd., 1956), pp. 167–169.]

[3] It goes without saying that work resumes in principle soon after the emission of sound. We still have the opportunity of reliving these original experiences in practical life. Thus the cry 'heave ho!' arises from the objective structure of a collective effort of traction and is immediately comprehensible in that very work. We know, moreover, that the verb occupies a privileged place in primitive languages.

5. REMARKS ON THE PRECEDING DEVELOPMENT

[1*] English translation, quoted from Ernst Fisher: *Marx in His Own Words*, translated by A. Bostock (Allen Lane, Penguin Books, Inc., 1968), Appendix, pp. 165–166.

6. USE-VALUE AND THE MOVEMENT OF SACRIFICE

[1] Spencer and Gillen: *Across Australia*, p. 198.
[2] Spencer and Gillen: *The Native Tribes of Australia* (New York: Dover Publishers, Inc., 1968), p. 9. See also: *Northern Tribes of Central Australia*, pp. 27–28.
[3] This can be corroborated in the custom of reopening the scars of the subincision in order to draw blood before leaving on a war-expedition. Cf. Strehlow: *Die Aranda und Loritja-Stamme in Zentral-Australien*, Vol. 4, Part II, p. 6.
[4] One will notice that many local groups can have the same totem which, obviously, makes no difference in regard to its essential function.
[5*] E. Fischer, (ed.) *Marx in His Own Words, op. cit.*, Appendix, pp. 165–166.

7. THE MOVEMENT OF WEALTH

[1] It is precisely such a structure that reappears when trade returns as the elementary form of barter. Thus Herodotus relates how the Carthaginians obtained gold in exchange for commodities in a certain country of the African coast: ". . . where they no sooner arrive but forthwith they unlade their wares, and, having disposed them after an orderly fashion along the beach, leave them, and, returning aboard their ships, raise a great smoke. The natives, when they see the smoke, come down to the shore, and laying out to view so much gold as they think the worth of the wares, withdraw to a distance. The Carthaginians upon this come ashore and look. If they think the gold enough, they take it and go their way; but if it does not seem to them sufficient, they go aboard ship once more, and wait patiently. Then the others approach and add to their gold, till the Carthaginians are content. Neither party deals unfairly with the other: for they them themselves never touch the gold till it comes up to the worth of their goods, nor do the natives ever carry off the goods till the gold is taken away." (IV, 196.) [English translation in: *The Histories of Herodotus*, translated by G. Rawlinson (New York: Modern Library, 1964), Volume 1, No. 196.]

8. MERCANTILE ECONOMY

[1] For a factual exposition, cf. Jacques Pirenne: *Histoire des institutions et du droit privé dans l'ancienne Égypte.*

9. MONETARY ECONOMY

[1] One will notice that the dialectic just described is valid only by virtue of a certain *rhythm* determined by the combination of objective conditions. With the Phoenicians, where mercantile economy had, for a long time, assured to urban life a high level of

development, the aristocracy had had the time to absorb new productive forces: the bourgeois revolution had no longer any *raison d'être*; and the primitive customs (such as human sacrifices) were retained until the Roman era.

2* English translation: J. Burnet, *Early Greek Philosophy* (New York: Meridian Books, 1957), p. 137.

3* *Ibid.*, p. 136.

4* *Ibid.*, p. 119.

5* *Ibid.*, pp. 174–175.

6* English translation: *Plato: The Collected Dialogues*, edited by E. Hamilton and H. Cairns (New York: Pantheon Books, 1961); *Protagoras*, translated by W. K. C. Guthrie, p. 323 (327d).

7* The translators have not been able to find this quotation in the *Protagoras.*

8* *Ibid.*, p. 300 (519a).

9* *Ibid.*, p. 956 (166b).

10. CAPITALISTIC ECONOMY

1* *Le Capital* (Éd. sociales), II, p. 50. [*Capital*, translated by Samuel Moore and Edward Aveling (New York: International Publishers, 1967), Volume 1, p. 361. We have extensively revised this translation by consulting the original German. – Tr.]

2* *Phénoménologie de l'Esprit*, in fine. [English translation, *op. cit.*, p. 808.]

3* *Manifeste communiste* (Éd. sociales), p. 31. [*The Communist Manifesto* (New York: International Publishers, 1968), p. 11.]

BIBLIOGRAPHY OF WORKS CITED

WORKS BY HUSSERL*

Logische Untersuchungen. Halle: Max Niemeyer Vol. I 1900, Voll II 1901. English translation: 1970. *Logical Investigations*. Tr. by J. N. Findlay from the 2nd edition of 1913/1921. London: Routledge and Kegan Paul.

Ideen I and II. English translation used: 1931. *Ideas*. Tr. by W. R. Boyce-Gibson. New York: Macmillan.

Formale und Transzendentale Logik. 1929. English translation: 1969. *Formal and Transcendental Logic*. Tr. by Dorian Cairns. The Hague: Martinus Nijhoff.

Erfahrung und Urteil. 1939 (ed. L. Landgrebe). English translation: 1973. J. S. Churchill and K. Ameriks. Evanston: Northwestern Univ. Press.

Méditations Cartésiennes. 1931. Translated by E. Levinas, from German original 1930. [Cf. Strasser edition of 1950.] English translation: 1960. *Cartesian Meditations*. Tr. by Dorian Cairns. The Hague: Martinus Nijhoff.

'Philosophie als Strenge Wissenschaft', *Logos* 1, 1911. 'Philosophy as Rigorous Science'. Tr. Quentin Lauer in *Phenomenology and the Crisis of Philosophy*. 1865. New York: Harper.

Philosophie der Arithmetik. 1891.

Die Krisis der Europaischen Wissenschaften und die Transzendentale Phänomenologie. 1936. English translation: 1970. *The Crisis of European Sciences and Transcendental Phenomenology*. Translated with an introduction by David Carr. Evanston: Northwestern University Press.

Die Idee der Phänomenologie. 1907. English translation: 1964. *The Idea of Phenomenology*. Translated by W. P. Alston and G. Nakhnikian. The Hague: Martinus Nijhoff.

'Vorlesungen zur Phänomenologie des inneren Zeitbewusstseins'. German text edited by M. Heidegger, 1928. English translation: 1964. *The Phenomenology of Internal Time-Consciousness*. Translated by J. S. Churchill. Bloomington: Indiana University Press.

OTHER WORKS CITED

Cavailles, Jean. *Méthode axiomatique et formalisme*. Paris: Hermann 1937.

Descartes, R. *Méditations*. English translation: 1931. 'Meditations on First Philosophy', in *The Philosophical Works of Descartes*. Translated by E. S. Haldane and G. R. T. Ross. New York: Dover.

Engels, F. *Ludwig Feuerbach et la fin de la philosophie classique allemande*, Vol. 2. Paris: Éditions sociales. English translation: 1967. *Ludwig Feuerbach and the Outcome of Classical German Philosophy*. New York: International Publishers.

* [Except where stipulated, Trân Duc Thao does not give publication data on his references to Husserl's works and those of other authors. – Translators' note.]

Engels, F. *Dialectique de la nature*. English translation: 1940. *Dialectics of Nature*. Edited and translated by C. Dutt. New York: International Publishers.

Fink, E. 1933. 'Die Phänomenologische Philosophie E. Husserls in der Gegenwartigen Kritik', *Kantstudien* 33.

Hegel, G. W. F. *Phénoménologie de l'esprit*. English translation: 1966. *The Phenomenology of Mind*. Translated by J. B. Baillie. New York.

Hering, J. 'La Phénoménologie d'Edmond Husserl il y a trente ans', *Revue internationale de philosophie* 2.

Hérodote. *Les histoires d'Hérodote*. English translation: 1964. *The Histories of Herodotus*. Translated by G. Rawlinson. New York: Modern Library.

Kant, I. *La critique de la raison pure*. English translation: 1961. *The Critique of Pure Reason*. Translated by N. K. Smith. New York: St. Martin Press.

Kant, I. *La critique du judgement*. English translation: 1964. *The Critique of Judgment*. Translated by James Meredith. Oxford: Clarendon Press.

Kant, I. *La religion dans les limites de la simple raison*. English translation: 1960. *Religion Within the Limits of Reason Alone*. New York: Harper.

Köhler, W. *L'intelligence des singes supérieurs*. Paris: Alcan. English original: 1956. *The Mentality of Apes*. London: Routledge and Kegan Paul.

Kreis, F. 1930. *Phänomenologie und Kritizismus*. Tübingen: Mohr .

Landgrebe. 'Husserls Phänomenologie und die Motive ihrer Umbildung', *Revue Internationale de Philosophie* 2.

Leibniz, G. W. von. *Nouveaux essais*. English translation: 1951. *'New Essays on Human Understanding'*, in *Leibniz, Selections*. New York: Scribner's.

Marx, K. *Contribution à la critique de l'économie politique*.

Marx, K. *Le capital*. Paris: Éditions sociales. English translation: 1967. *Capital*. Translated by S. Moore and Ed. Aveling. New York: International Publishers.

Marx K. and F. Engels. *Manifeste communiste*. Paris: Éditions sociales. English translation: 1968. *The Communist Manifesto*. New York: International Publishers.

Natorp, Paul. 'Husserls Ideen zu einer reinen Phänomenologie', *Logos* 7.

Pirenne, J. 1982–1985. Histoire des institutions et du droit privé dans l'ancienne Égypte. Brussels.

Platon. Gorgias, Protagoras. *Parménide*, English translation in *Plato, The Collected Dialogues*. Edited by E. Hamilton and H. Cairns. Bollingen Series, Pantheon Books.

Rickert, H. 'Die Methode der Philosophie und das Unmittelbare', *Logos* 12.

Rickert, H. 'Vom Anfang der Philosophie', *Logos* 14.

Spencer, B. and F. J. Gillen 1899. *Across Australia*. London: Macmillan (2 vols.), 1968.

Spencer, B. and F. J. Gillen. 1968. *The Native Tribes of Central Australia*. New York: Dover.

Spencer, B. and F. J. Gillen 1904. *Northern Tribes of Central Australia*. Oosterhout: Anthropol. Pub. 1969.

Spinoza, B. *Éthique*. English translation: 1951. *The Ethics*. Translated with an introduction by R. H. M. Elwes. New York: Dover Publications, Inc.

Strehlow, C. 1907–20 *Die Aranda und Loritja-Stamme in Zentral-Australien*. Frankfurt-a-M.: Baer.

Zocher, R. 1932. *Husserls Phanomenologie und Schuppes Logik*, Munich: Reinhardt.

INDEX OF NAMES

9 789027 707376

GPSR Compliance

The European Union's (EU) General Product Safety Regulation (GPSR) is a set of rules that requires consumer products to be safe and our obligations to ensure this.

If you have any concerns about our products, you can contact us on ProductSafety@springernature.com

In case Publisher is established outside the EU, the EU authorized representative is:

Springer Nature Customer Service Center GmbH
Europaplatz 3
69115 Heidelberg, Germany

Batch number: 10755292

Printed by Printforce, the Netherlands